KEY TO GEOMETRY

Book 7: TABLE OF CONTENTS

TO THE STUDENT:

These books will help you to discover for yourself many important relationships of geometry. Your tools will be the same as those used by the Greek mathematicians more than 2000 years ago. These tools are a **compass** and a **straightedge**. In addition, you will need a **sharpened pencil**. The lessons that follow will help you make drawings from which you may learn the most.

The answer books show **one** way the pages may be completed correctly. It is possible that your work is correct even though it is different. If your answer differs, re-read the instructions to make sure you followed them step by step. If you did, you are probably correct.

Scholars of the Arab world have played an important role in the history of mathematics. About 800 A.D. Moslem rulers in Baghdad set out to make the city a great center of learning, a second Alexandria. They sent representatives to travel throughout the known world, collecting works by the great mathematicians from both East and West. For centuries Islamic scholars translated Greek, Hindu and Chinese works into Arabic. Among the most popular of the works at Baghdad was the **Elements** by Euclid. About 1000 A.D. Arab scholars brought their knowledge of geometry to Europe, establishing famous schools that helped rekindle European interest in mathematics.

On the cover of this booklet Moslem workmen tile a wall of a mosque. Because the prophet Mohammed had forbidden the use of human or animal forms in Islamic art, Moslems based their art on repeating geometric designs called tesselations. All of the complicated tesselation patterns used in Islamic art were created with the classic tools of geometry, the compass and the straightedge.

Published by Key Curriculum Press, 1150 65th Street, Emeryville, CA 94608
Printed in the United States of America
10 LKV 21
ISBN 978-0-913684-77-1

Comparing Angles

Problem: *Compare angles.*

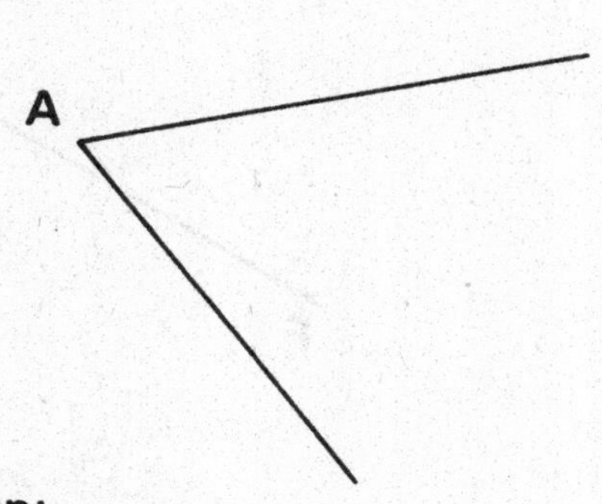

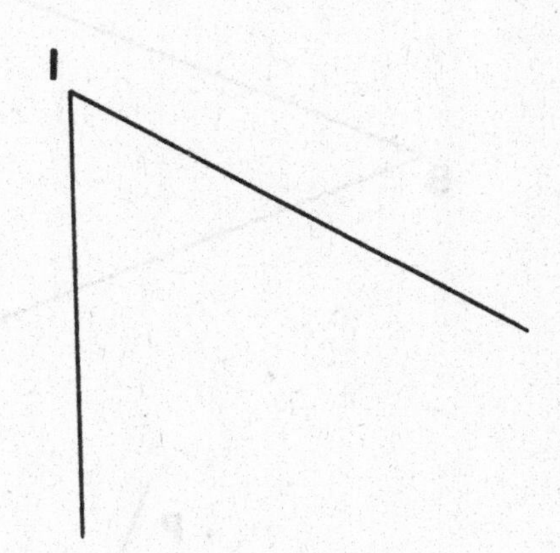

Solution:

1. Draw an arc with center A intersecting both sides of the angle.
   Label the points of intersection B and G.

2. Use the same radius and center I to draw an arc intersecting both sides of the other angle.
   Label the points of intersection V and P.

3. Compare $\overline{BG}$ and $\overline{VP}$.

   $\overline{BG}$ is _ _ _ _ _ _ _ _ _ _ _ _ _ _ _ _ _ _ $\overline{VP}$.

   (a) shorter than (c) longer than

   (b) congruent to

4. Angle BAG is _ _ _ _ _ _ _ _ _ _ _ _ _ _ _ _ _ _ angle VIP.

   (a) smaller than (c) larger than

   (b) congruent to

5. Compare angle XYZ and angle UVW.

   Angle XYZ is _ _ _ _ _ _ _ _ _ _ _ _ angle UVW.

   (a) smaller than (c) larger than

   (b) congruent to

X

Y

Z

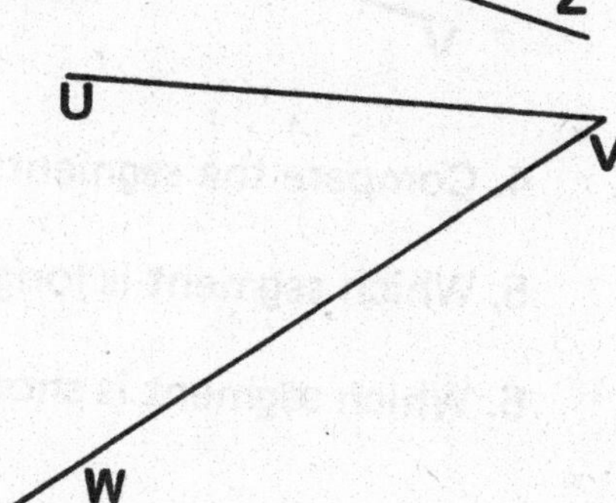

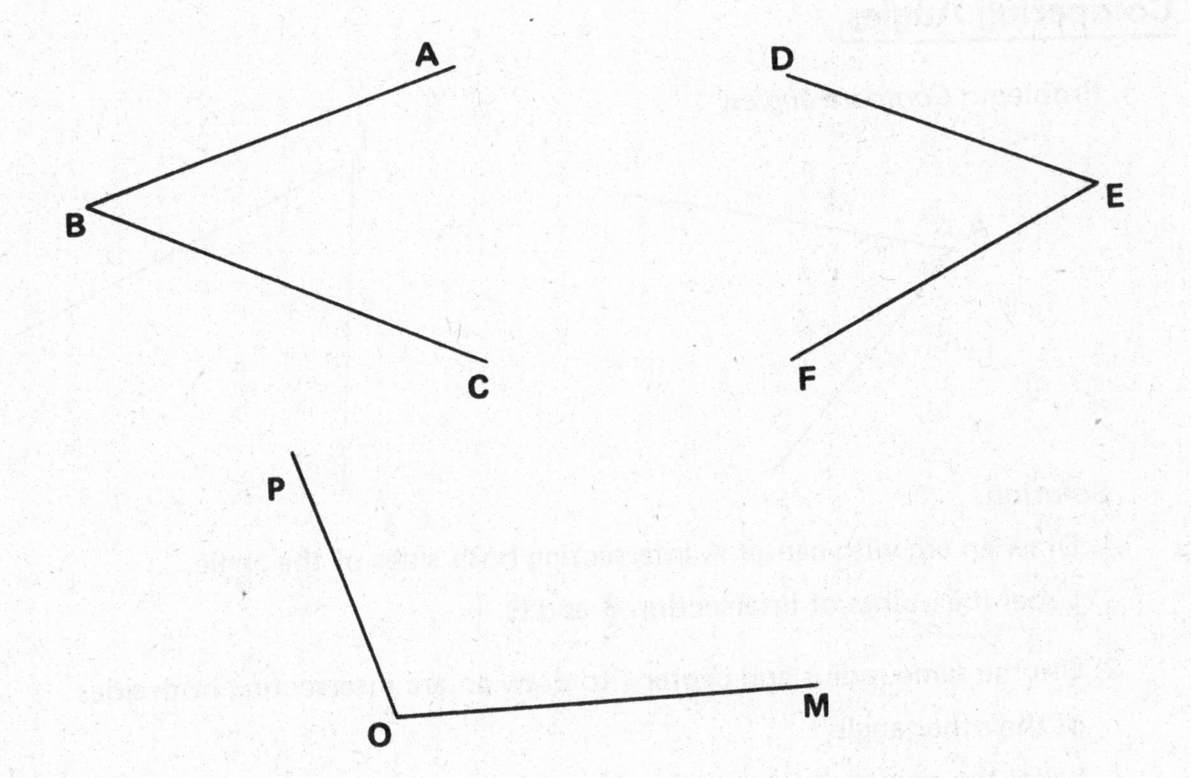

1. Compare the angles.

2. Which angle is largest? _ _ _ _ _ _

3. Which angle is smallest? _ _ _ _ _ _

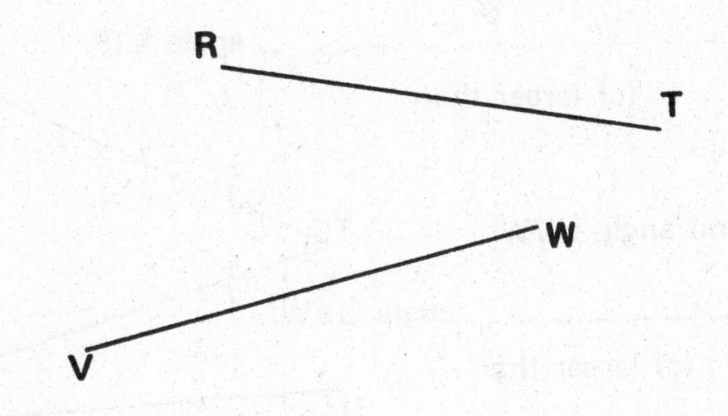

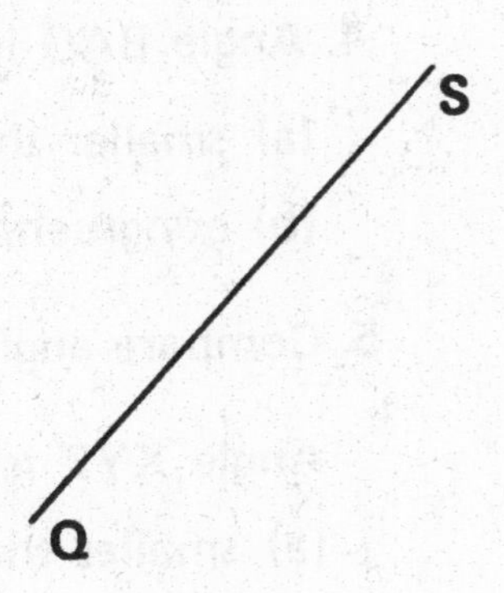

4. Compare the segments.

5. Which segment is longest? _ _ _ _ _ _

6. Which segment is shortest? _ _ _ _ _ _

Vertical Angles

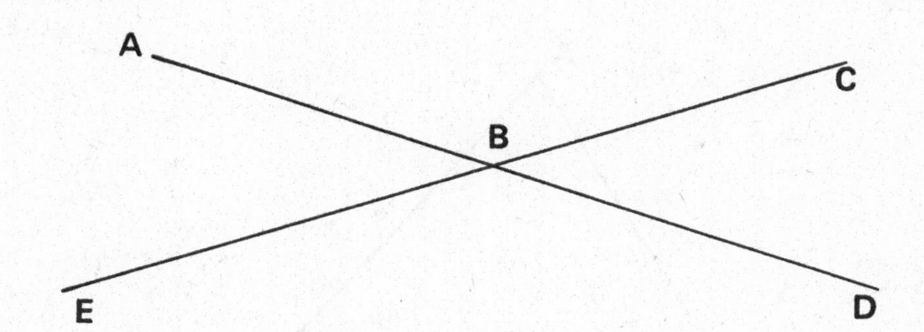

1. Angle ABE and angle CBD are vertical angles.
   Compare angle ABE and angle CBD.

   Angle ABE is _ _ _ _ _ _ _ _ _ _ _ _ _ _ _ _ _ _ angle CBD.

   (a) smaller than (c) larger than

   (b) congruent to

2. Angle ABC and angle EBD are also vertical angles.
   Compare angle ABC and angle EBD.

   Angle ABC is _ _ _ _ _ _ _ _ _ _ _ _ _ _ _ _ _ _ angle EBD.

   (a) smaller than (c) larger than

   (b) congruent to

P
•

3. Draw two lines through point P.

4. Label a pair of vertical angles as angle QPR and angle SPT.

5. Compare vertical angles QPR and SPT.

   Are the vertical angles congruent? _ _ _ _ _

The Sides and Angles of a Triangle

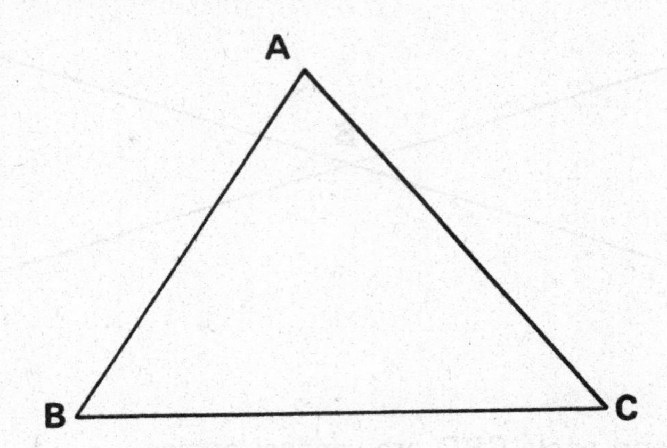

1. Compare the three sides of triangle ABC.

   ______ is the longest side.

   ______ is the shortest side.

2. Compare the three angles of triangle ABC.

   ______ is the largest angle.

   ______ is the smallest angle.

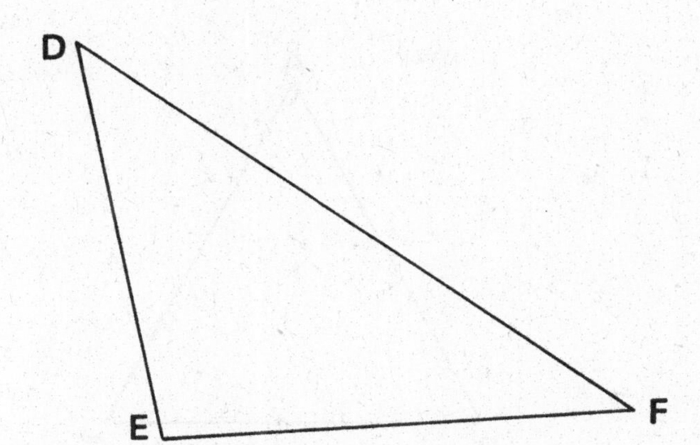

1. Compare the sides and angles of triangle DEF.

_ _ _ _ _ _ is the longest side and _ _ _ _ _ _ is the largest angle.

_ _ _ _ _ _ is the shortest side and _ _ _ _ _ _ is the smallest angle.

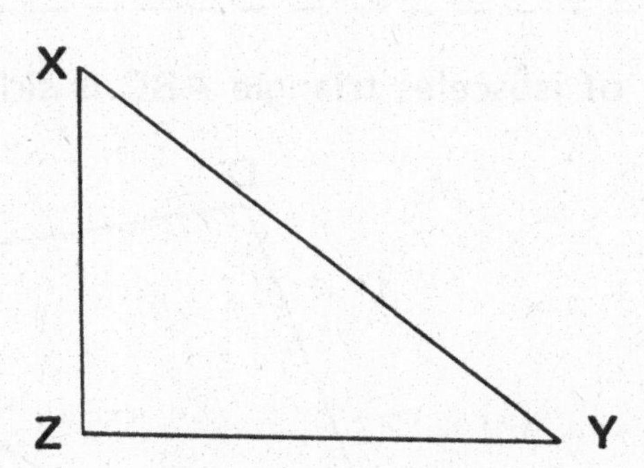

2. Answer these questions without using your compass.

$\overline{XY}$ is the longest side and _ _ _ _ _ _ is the largest angle.

_ _ _ _ _ _ is the shortest side and angle XYZ is the smallest angle.

The Isosceles Triangle

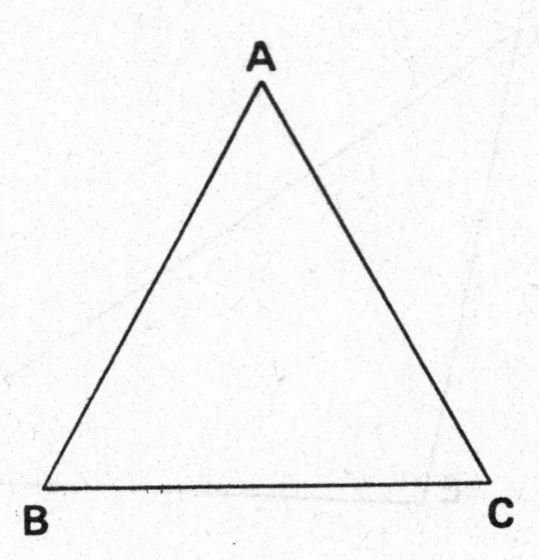

1. Compare sides $\overline{AB}$ and $\overline{AC}$.

   Are they congruent? _ _ _ _ _

2. A triangle with two congruent sides is an isosceles triangle.

   Is triangle ABC an isosceles triangle? _ _ _ _ _

   Why? _

3. The base of isosceles triangle ABC is side $\overline{BC}$.

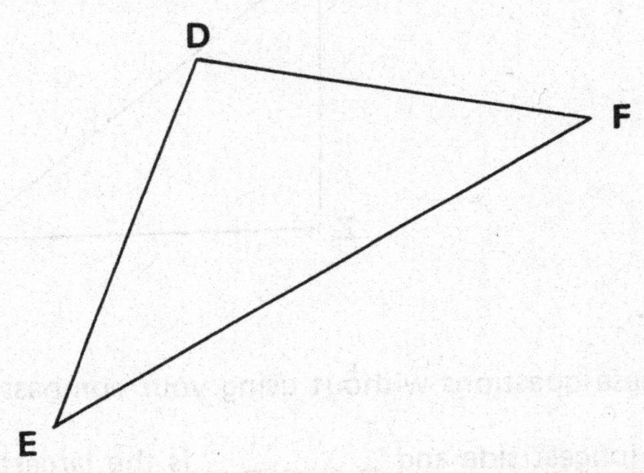

4. Check with your compass. Is triangle DEF an isoceles triangle? _ _ _

   Why? _

5. Which side is the base of triangle DEF? _ _ _ _ _

Problem: *Construct an isosceles triangle whose base is $\overline{XY}$.*

Solution: X ———————— Y

1. Draw an arc with center X and any radius more than half of $\overline{XY}$.
2. With the same radius, draw an arc with center Y.
   Make the arcs intersect.
3. Label the point of intersection Z.
4. Draw triangle XYZ.
5. Is triangle XYZ an isosceles triangle? _ _ _ _ _
   Why? _
6. Construct another isosceles triangle with base $\overline{PQ}$.

P

Q

7. How many isosceles triangles can you draw with base $\overline{PQ}$? _ _ _ _ _

1. Draw a circle with center Q.

Q •

2. Draw two radii of the circle and label them $\overline{QR}$ and $\overline{QS}$.

3. Draw triangle QRS.

4. Is triangle QRS isosceles? _ _ _ _ _

   Why? _

5. Name the base of triangle QRS. _ _ _ _ _

6. Construct an isosceles triangle below.

1. Is triangle GEF an isosceles triangle? _ _ _ _ _

Why? _

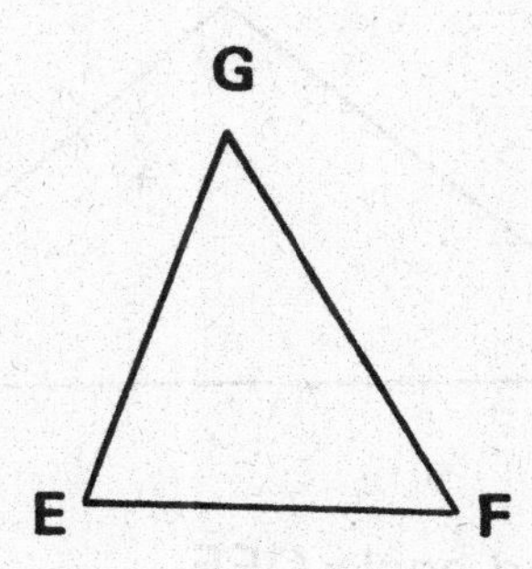

2. Which triangle is isosceles? _ _ _ _ _

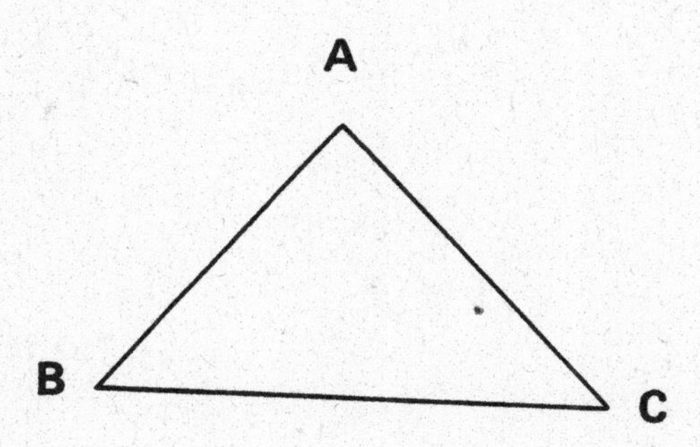

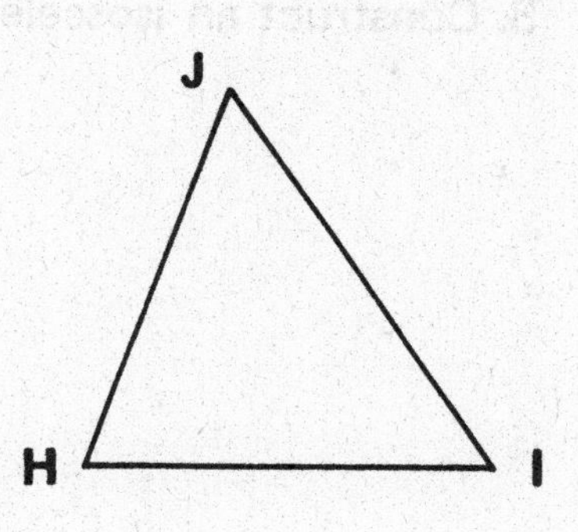

Name the base of the isosceles triangle. _ _ _ _ _

3. Construct two isosceles triangles with $\overline{MN}$ as base.

1. Is triangle DEF an isosceles triangle? _ _ _ _ _

   Why? _

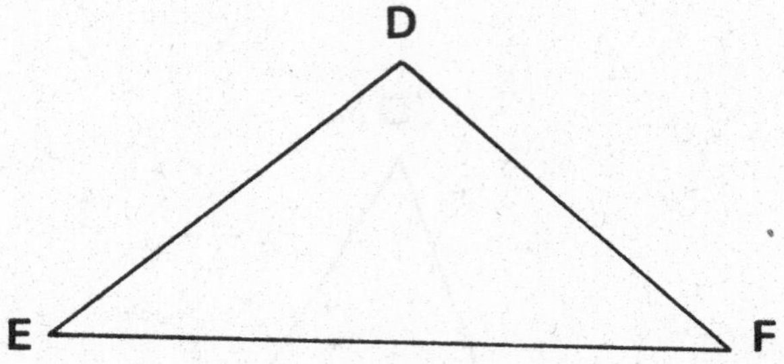

D

E F

2. Compare angle DEF and angle DFE.

   Are they congruent? _ _ _ _ _

3. Construct an isosceles triangle.

   Label it MNP.

4. Name the two congruent sides in triangle MNP. _ _ _ _ _

5. Name the base of triangle MNP. _ _ _ _ _

6. Compare the three angles in triangle MNP.

   Name two congruent angles. _ _ _ _ _

7. An isosceles triangle has two congruent sides and

   _

Review

1. Compare the angles.

   Are they congruent? _ _ _ _ _

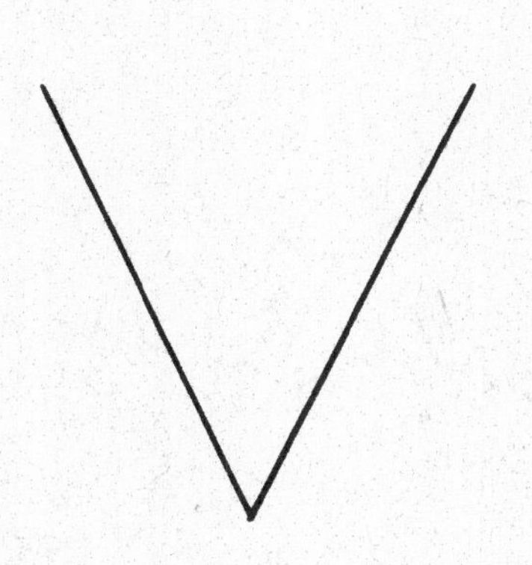

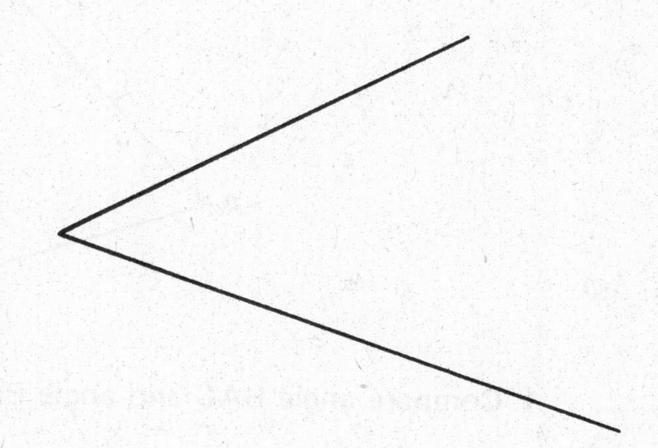

2. Construct an angle, with vertex A and ray $\overrightarrow{AB}$ as one side, which is congruent to angle DEF.

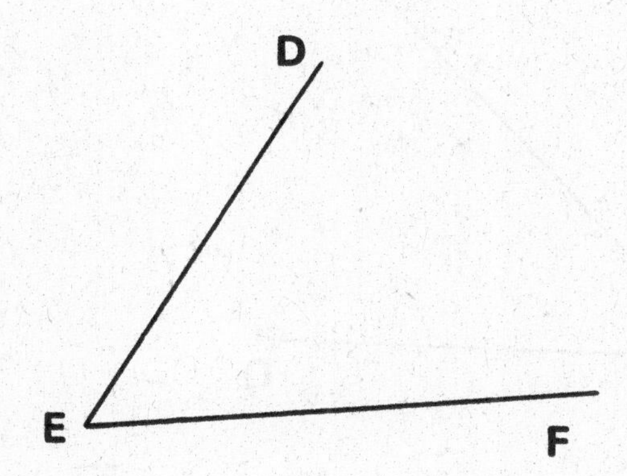

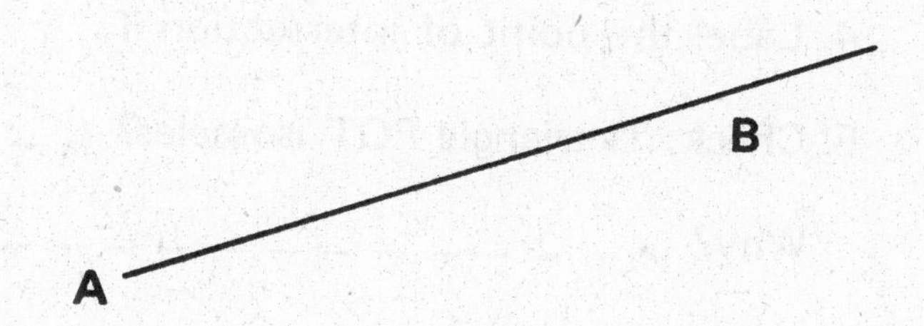

The Isosceles Triangle

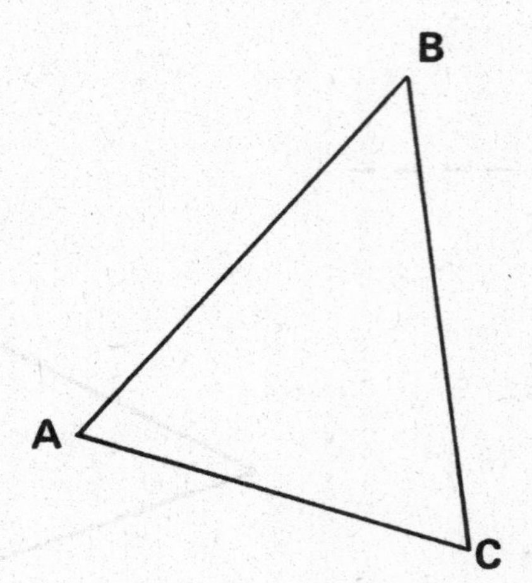

1. Compare angle BAC and angle BCA.

   Are they congruent? _ _ _ _ _

2. Is triangle ABC an isosceles triangle? _ _ _ _ _

   Name two congruent sides. _ _ _ _ _ _ _ _ _ _ _ _ _ _ _ _

   Name its base. _ _ _ _ _

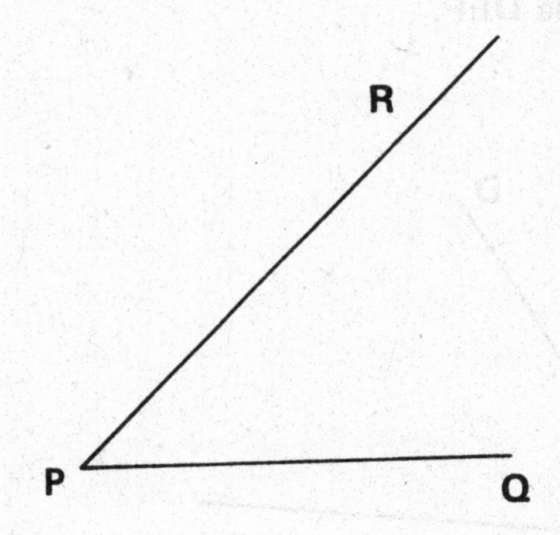

3. At Q construct an angle congruent to angle RPQ with $\overleftrightarrow{PQ}$ as one side. Make the other side intersect $\overrightarrow{PR}$.

4. Label the point of intersection T.

5. Check: Is triangle PQT isosceles? _ _ _ _ _

   Why? _

The Perpendicular Bisector of a Segment

Problem: *Construct the perpendicular bisector of a segment.*

A ———————————— B

Solution:

1. Draw an arc with center A and a radius more than half $\overline{AB}$.
2. Draw an arc with center B and congruent radius.
   Make the arcs intersect in two points.
3. Label the intersections C and D.
4. Draw $\overleftrightarrow{CD}$.
5. $\overleftrightarrow{CD}$ is _ _ _ _ _ _ _ _ _ _ _ _ _ _ _ _ _ _ $\overline{AB}$.

   (a) the right angle of
   (b) congruent to
   (c) the perpendicular bisector of
   (d) the diagonal of

6. Label as M the midpoint of $\overline{AB}$.

1. Construct the perpendicular bisector of each segment.

2. Construct an equilateral triangle with $\overline{AB}$ as one side.

The Perpendicular Bisector of the Base of a Triangle

1. Construct the perpendicular bisector of $\overline{BC}$.

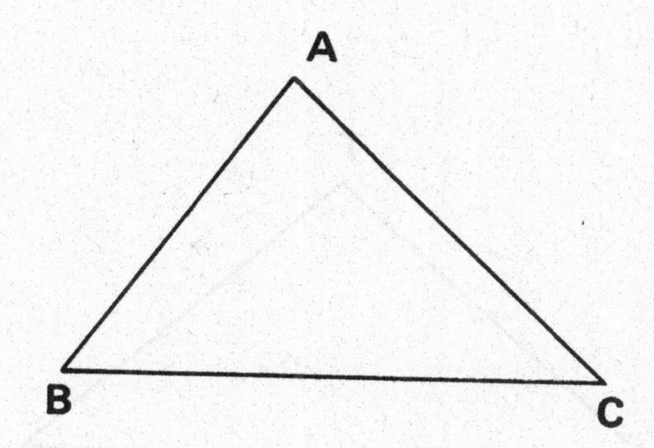

2. Does the perpendicular bisector pass through the opposite vertex A? _ _ _ _ _

3. Check: Is triangle ABC an isosceles triangle? _ _ _ _ _

   Why? _

4. Construct an equilateral triangle.

5. Construct the perpendicular bisector of one side.

6. Does it pass through the opposite vertex? _ _ _ _ _

7. Is the equilateral triangle an isosceles triangle? _ _ _ _ _

   Why? _

8. The perpendicular bisector of a side of a triangle _ _ _ _ _ _ _ _ _ _ passes through the opposite vertex.

   (a) always (b) sometimes (c) never

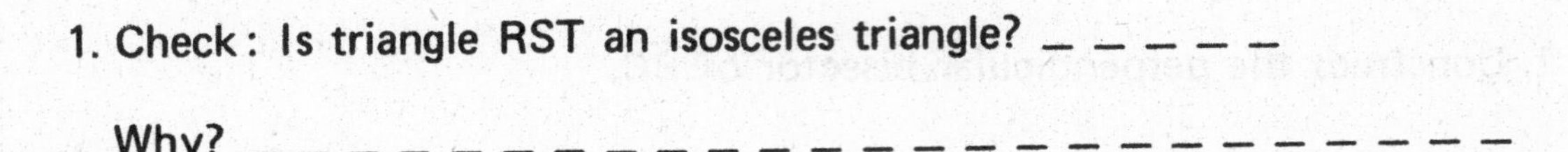

1. Check: Is triangle RST an isosceles triangle? _____

   Why? ______________________________

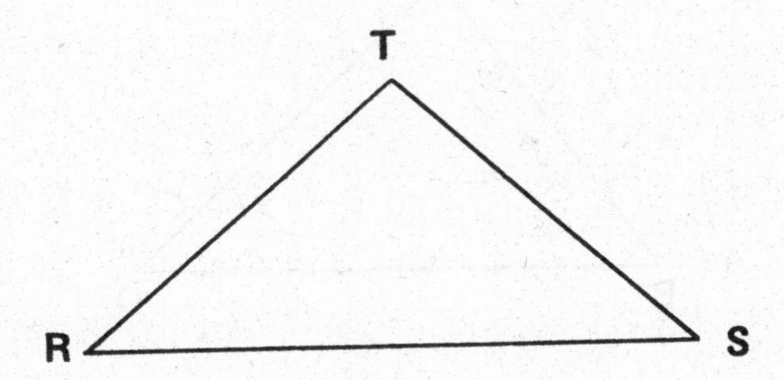

2. Construct the perpendicular bisector of base $\overline{RS}$.

   Does it pass through the opposite vertex T? _____

3. Construct an isosceles triangle.

4. Construct the perpendicular bisector of the base.

   Does it pass through the opposite vertex? _____

5. The perpendicular bisector of the base of an isosceles triangle

   _________________ passes through the opposite vertex.

   (a) always (b) sometimes (c) never

The Bisector of an Angle

Problem: *Bisect an angle.*

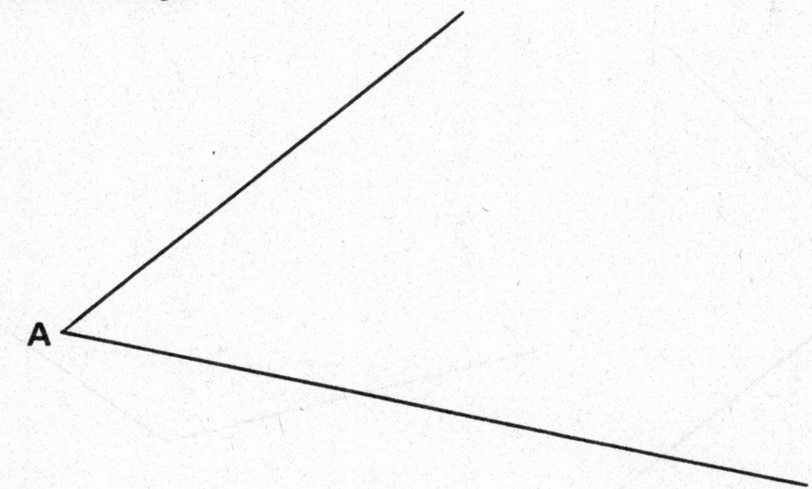

Solution:

1. Draw an arc with center A which intersects both sides of the angle.

2. Label the intersections as B and D.

3. Draw an arc with B as center.

   Use the same radius to draw an arc with D as center.

   Make these arcs intersect.

4. Label the intersection C.

5. Draw $\overrightarrow{AC}$.

6. Compare angle BAC and angle CAD.

   Are they congruent? _ _ _ _ _

7. $\overrightarrow{AC}$ is _ _ _ _ _ _ _ _ _ _ _ _ _ _ _ _ _ _ angle BAD.

   (a) congruent to
   (b) the bisector of
   (c) perpendicular to
   (d) the midpoint of

1. Bisect each angle.

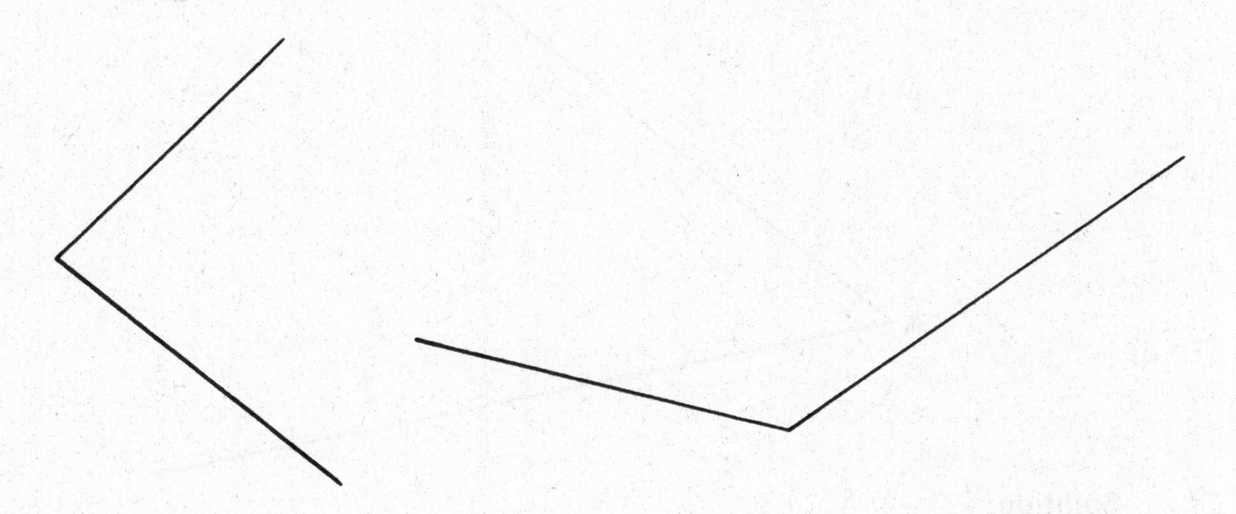

2. Bisect the straight angle.

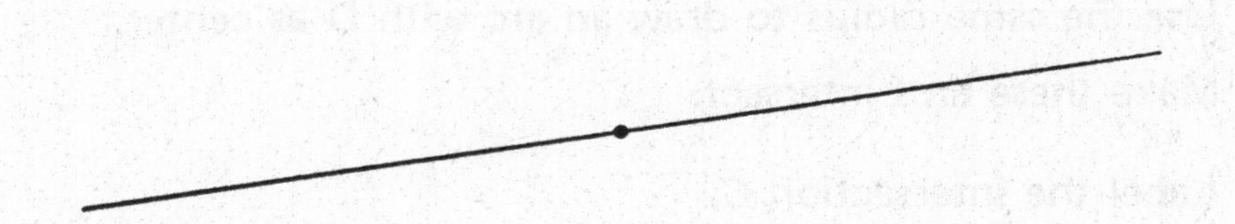

3. Check: Is M the midpoint of the segment? _ _ _ _ _

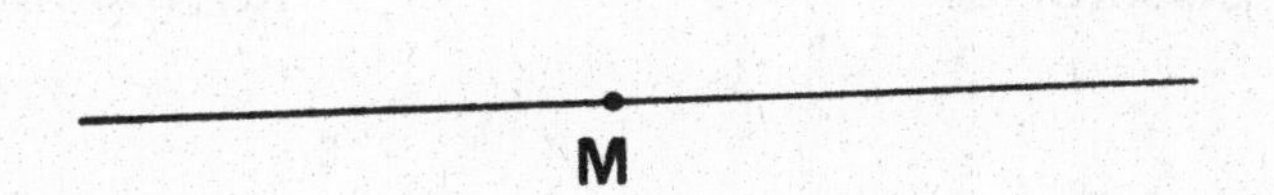

The Bisector of an Angle of a Triangle

1. Is triangle ABC an isosceles triangle? _ _ _ _ _

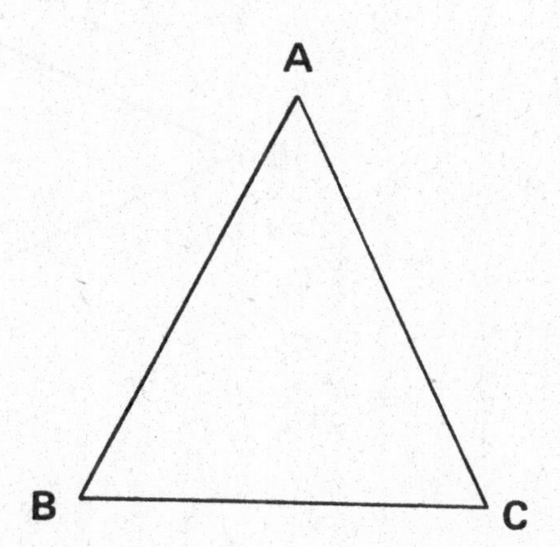

2. Construct the perpendicular bisector of $\overline{BC}$. Label as M the midpoint of $\overline{BC}$.

3. Compare angle BAM and angle CAM.

   Angle BAM is _ _ _ _ _ _ _ _ _ _ _ _ _ _ _ _ _ angle CAM.

   (a) smaller than (c) larger than

   (b) congruent to

4. $\overrightarrow{AM}$ _ _ _ _ _ _ _ _ _ _ _ _ _ _ _ _ _ _ angle BAC.

   (a) is congruent to (c) doubles

   (b) bisects (d) is perpendicular to

5. The perpendicular bisector of the base of an isosceles triangle

   _ _ _ _ _ _ _ _ _ _ _ _ _ _ _ _ _ bisects the opposite angle.

   (a) always (b) sometimes (c) never

1. Is triangle ABC isosceles? _ _ _ _ _

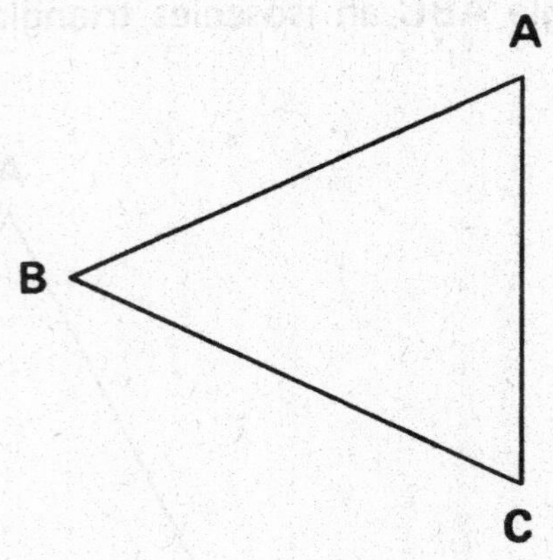

2. Bisect angle ABC.

3. Label as D the point where the bisector intersects $\overleftrightarrow{AC}$.

4. Check: Is AD congruent to DC? _ _ _ _ _

5. D is the _ _ _ _ _ _ _ _ _ _ _ _ _ _ _ _ _ of $\overline{AC}$.

   (a) midpoint (c) endpoint

   (b) vertex

6. Construct an isosceles triangle with base $\overline{XY}$.

   Label the other vertex Z.

X Y

7. Bisect angle XZY.

   Label as W the point where the bisector intersects $\overline{XY}$.

8. Check: Is W the midpoint of $\overline{XY}$? _ _ _ _ _

1. Is triangle DEF isosceles? _ _ _ _ _

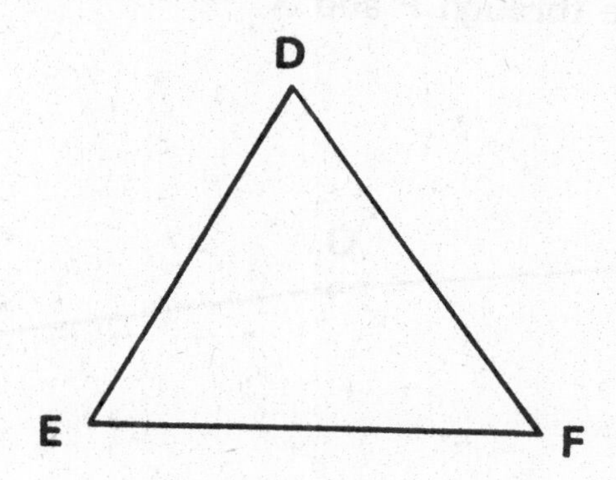

2. Bisect angle EDF.

   Label as G the point where it intersects $\overline{EF}$.

3. Is G the midpoint of $\overline{EF}$? _ _ _ _ _

4. Construct an equilateral triangle.

5. Bisect one of the angles of the equilateral triangle.

6. Does the bisector of the angle pass through the midpoint of the opposite side? _ _ _ _ _

7. The bisector of an angle of a triangle _ _ _ _ _ _ _ _ _ _ _ _ _ _ _ passes through the midpoint of the opposite side.

   (a) always (b) sometimes (c) never

8. When will the bisector on an angle of a triangle pass through the midpoint of the opposite side?

   _

Perpendicular Lines

1. Draw a straight line through P and X.

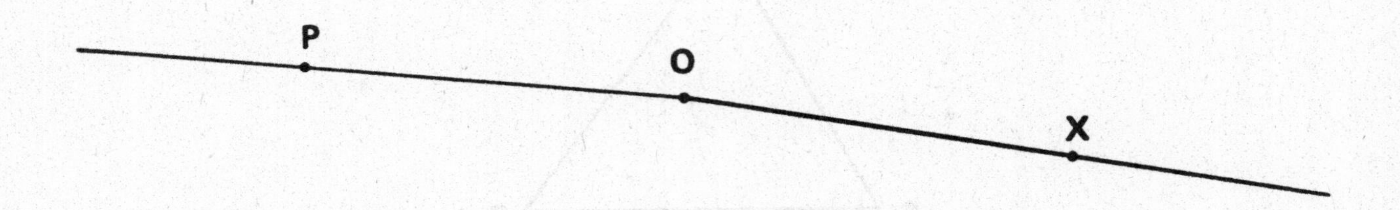

2. Does it pass through O? _ _ _ _ _

3. A straight angle is an angle whose sides form a straight line.

   Is angle POX a straight angle? _ _ _ _ _

4. Which of the angles below is a straight angle? _ _ _ _ _

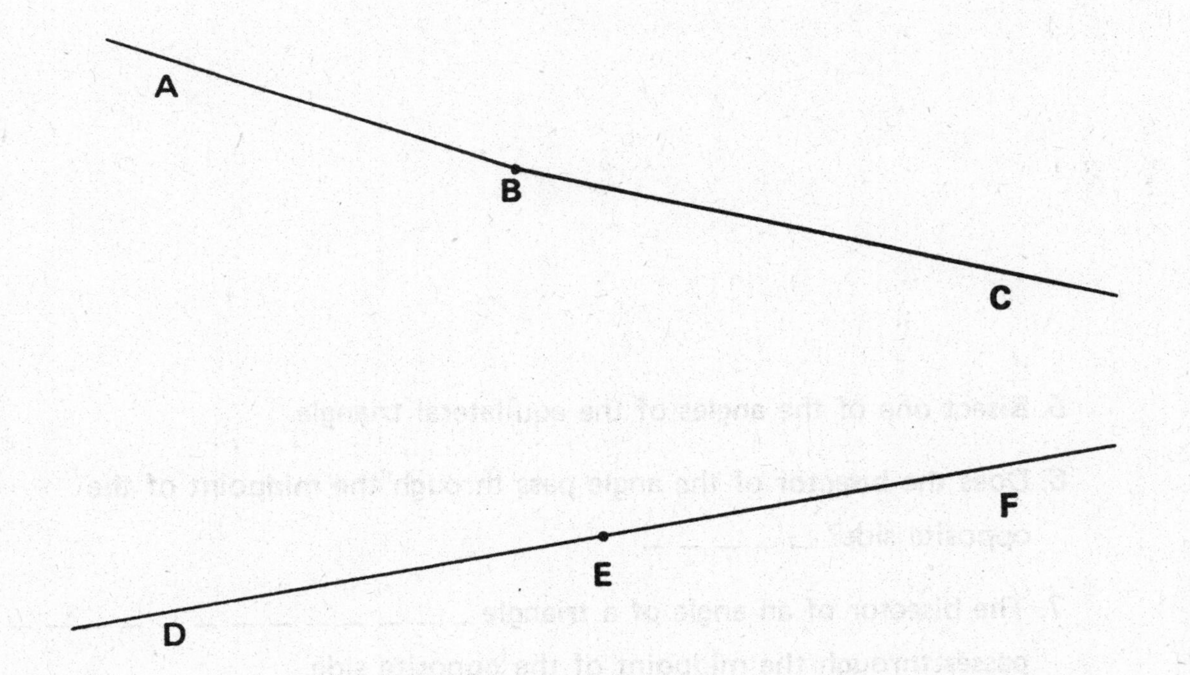

1. Is angle ABC a straight angle? _ _ _ _ _

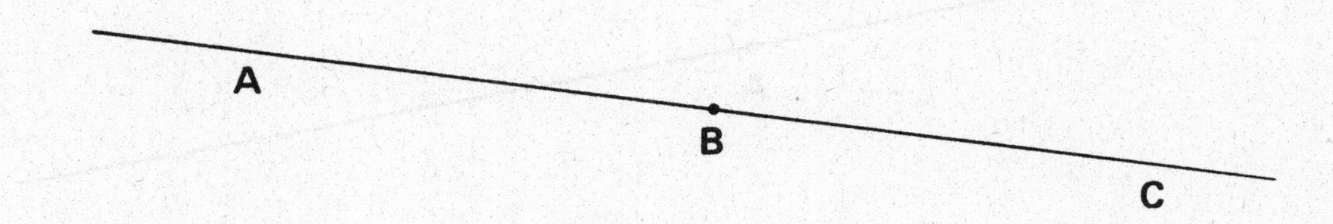

2. Bisect angle ABC.

3. Label as D a point on the bisector.

4. Line $\overleftrightarrow{BD}$ is _ _ _ _ _ _ _ _ _ _ _ _ _ _ _ _ _ line $\overleftrightarrow{AC}$.

(a) parallel to

(b) the bisector of

(c) perpendicular to

5. Bisect the straight angle below.

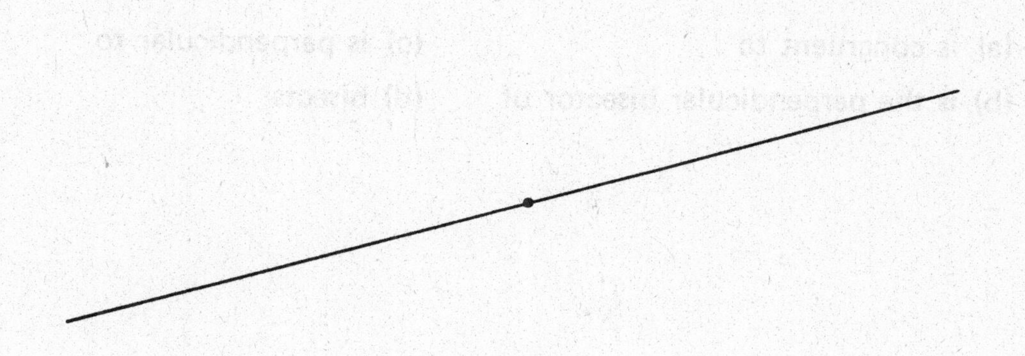

Problem: *Construct a line perpendicular to the given line through the point on the line.*

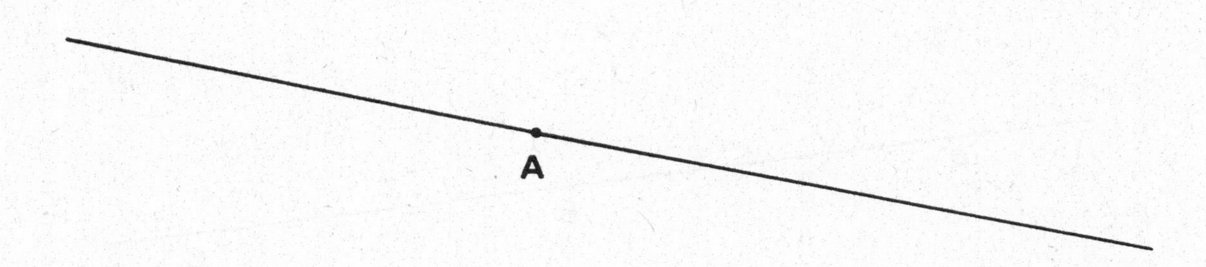

Solution:

1. Draw two arcs with A as center and the same radius which intersect the given line.
2. Label the intersections B and C.
3. Draw an arc with B as center and radius larger than AB above the line.
4. Draw an arc with C as center and the same radius.
   Make the arcs intersect.
5. Label as D the intersection.
6. Draw $\overleftrightarrow{AD}$.
7. $\overleftrightarrow{AD}$ _ _ _ _ _ _ _ _ _ _ _ _ _ _ _ _ _ line $\overleftrightarrow{BC}$.

   (a) is congruent to
   (b) is the perpendicular bisector of
   (c) is perpendicular to
   (d) bisects

1. Construct the perpendicular to each line through the given point.

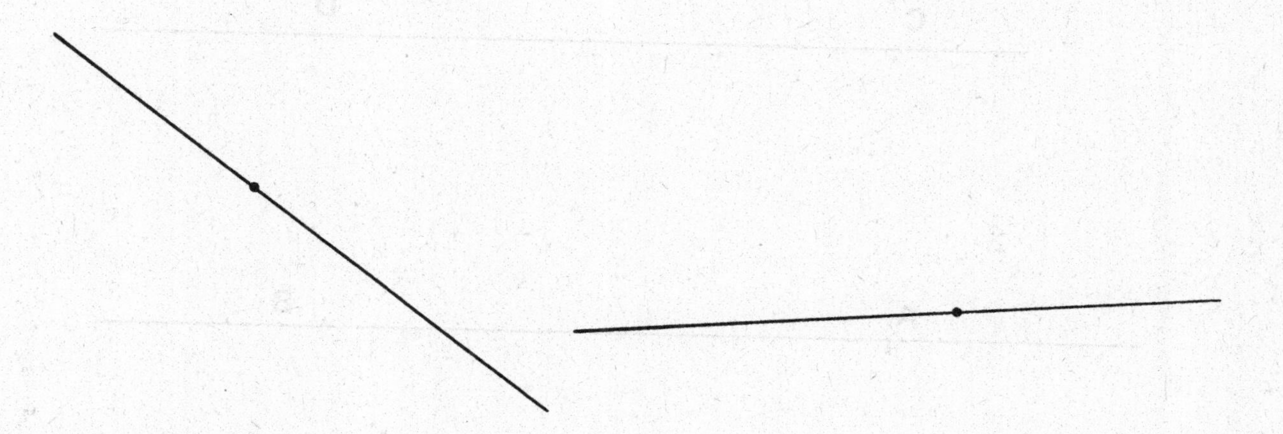

2. Check: Are the lines perpendicular? _ _ _ _ _

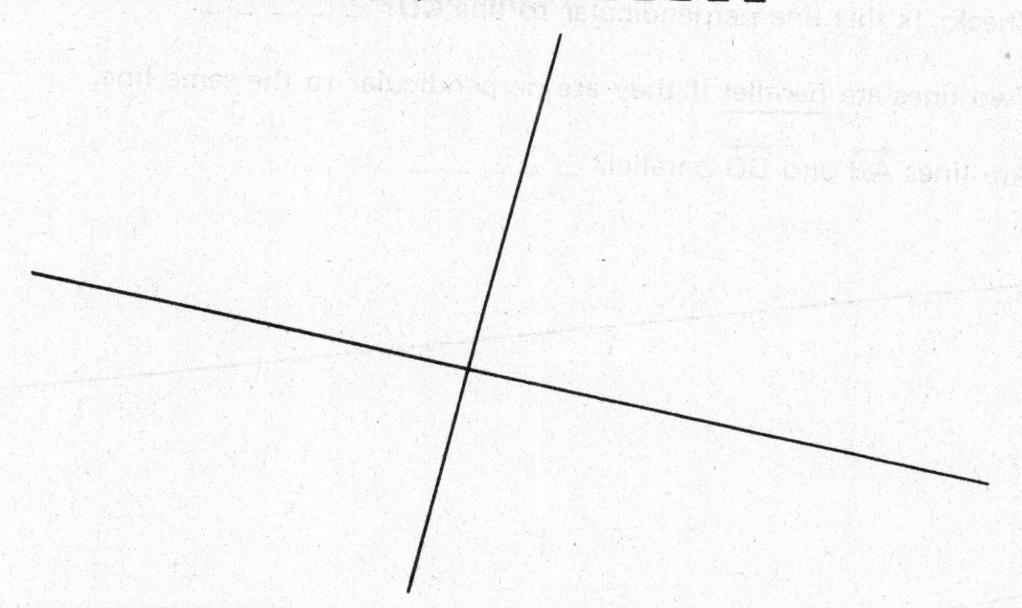

3. Construct a perpendicular to the given line.

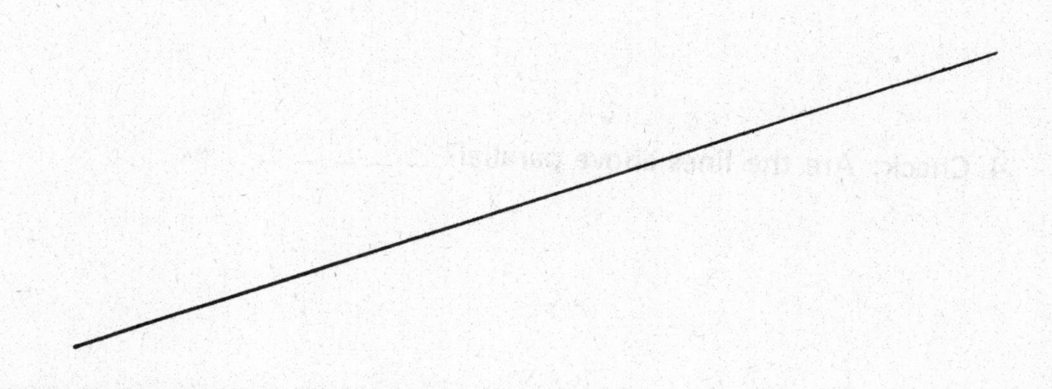

Parallel Lines

C D

A B

1. Construct a perpendicular to line $\overleftrightarrow{AB}$.

2. Check: Is this line perpendicular to line $\overleftrightarrow{CD}$? _ _ _ _ _

3. Two lines are parallel if they are perpendicular to the same line.

   Are lines $\overleftrightarrow{AB}$ and $\overleftrightarrow{CD}$ parallel? _ _ _ _ _

4. Check: Are the lines above parallel? _ _ _ _ _

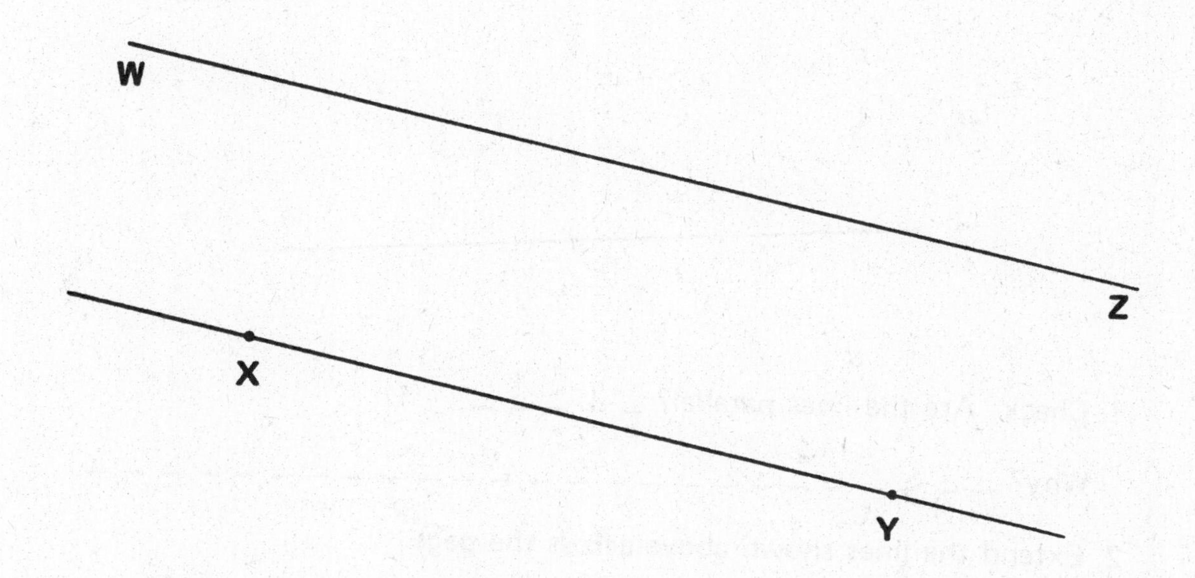

1. Construct a perpendicular to line $\overleftrightarrow{XY}$ at X.

2. Check: Is this line perpendicular to $\overleftrightarrow{WZ}$? _ _ _ _ _ _

3. $\overleftrightarrow{WZ}$ is _ _ _ _ _ _ _ _ _ _ _ _ _ _ _ _ _ _ $\overleftrightarrow{XY}$.

   (a) congruent to
   (b) perpendicular to
   (c) parallel to
   (d) the bisector of

4. Construct a perpendicular to line $\overleftrightarrow{XY}$ at Y.

5. Check: Is this line perpendicular to $\overleftrightarrow{WZ}$? _ _ _ _ _ _

1. Check: Are the lines parallel? _ _ _ _ _

Why? _

2. Extend the lines shown above across the page.

Will they intersect? _ _ _ _ _

3. Check: Are the lines parallel? _ _ _ _ _

Why? _

4. Extend the lines shown above.

Will they intersect? _ _ _ _ _

5. Two parallel lines will _ _ _ _ _ _ _ _ intersect.

(a) always (b) sometimes (c) never

The Perpendicular to a Line from a Point not on the Line

Problem: *Construct a line perpendicular to the given line through the given point.*

P •

A B

Solution:

1. Draw an arc with center P which intersects $\overleftrightarrow{AB}$ in two points..
2. Label the points of intersection as R and S.
3. Draw an arc with R as center below line $\overleftrightarrow{AB}$.
4. Draw an arc with S as center and the same radius.
   Make the arcs intersect.
5. Label their intersection T.
6. Draw $\overleftrightarrow{PT}$.
7. Is $\overleftrightarrow{PT}$ perpendicular to line $\overleftrightarrow{AB}$? _ _ _ _ _

1. Construct the perpendicular to each line through the given point.

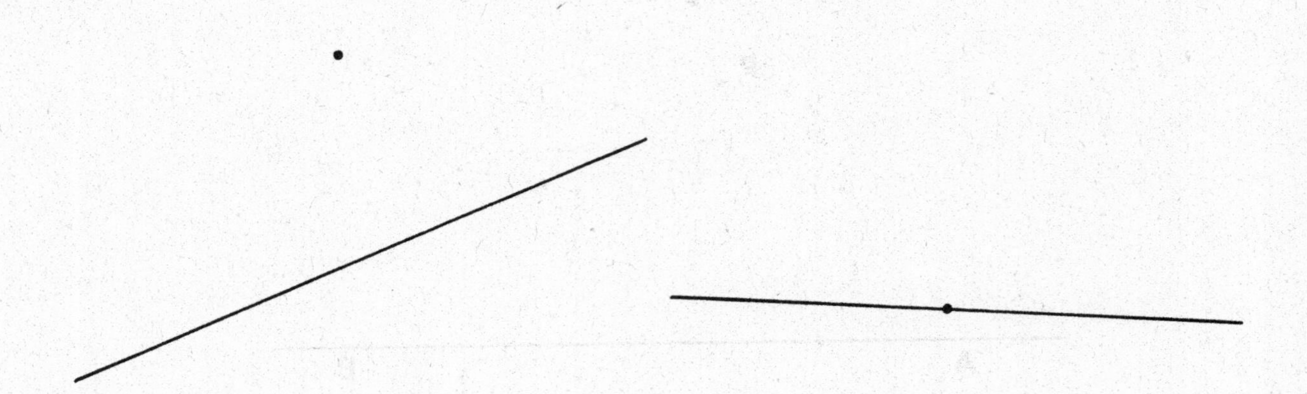

2. Check: Are the lines perpendicular? _ _ _ _ _

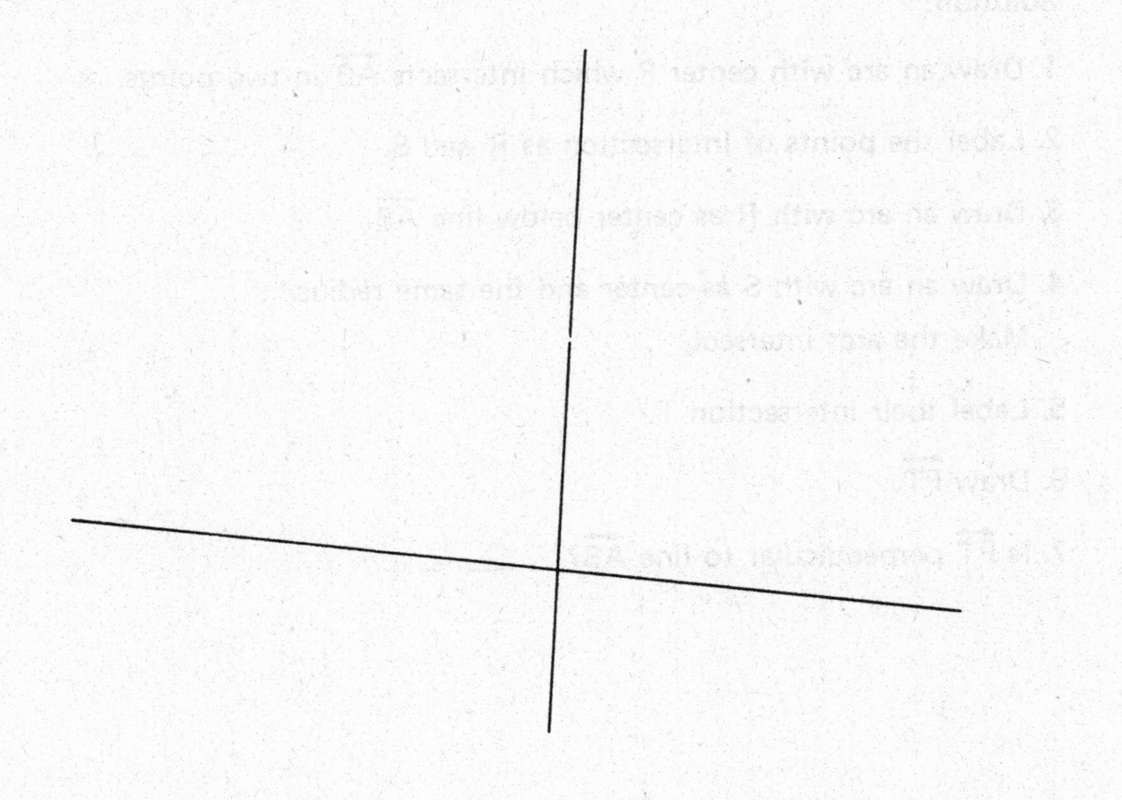

Constructing Parallel Lines

Problem: *Construct a line parallel to a given line through a given point.*

P

X

Y

Solution:

1. Construct a perpendicular to $\overleftrightarrow{XY}$ through P.

2. Label as Q the intersection of the perpendicular and $\overleftrightarrow{XY}$.

3. Construct a perpendicular to $\overleftrightarrow{QP}$ through P.

4. Is this line parallel to $\overleftrightarrow{XY}$? _ _ _ _ _

   Why? _

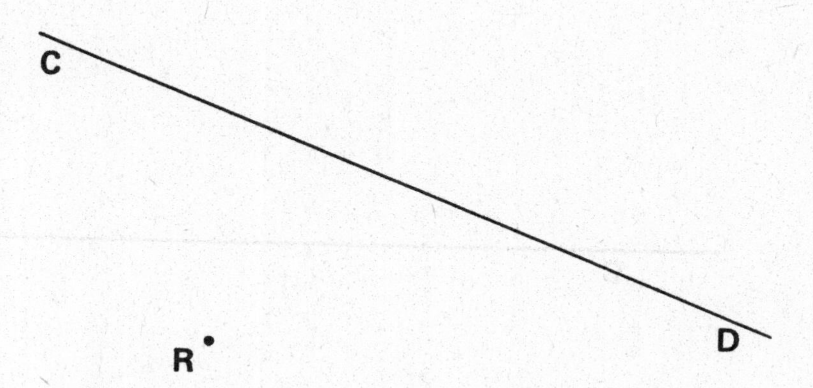

5. Construct a line parallel to $\overleftrightarrow{CD}$ through point R.

1. Construct a line perpendicular to $\overleftrightarrow{AB}$ through point R.

R

A

B

2. Construct a line parallel to line $\overleftrightarrow{AB}$ through point R.

3. Construct a line parallel to $\overleftrightarrow{CD}$ through point Q.

Q

C

D

1. Construct a perpendicular to line $\overleftrightarrow{MN}$ through point M.

Q P

N M

2. Label as K the point where the perpendicular intersects line $\overleftrightarrow{QP}$.

3. Check: Is this line perpendicular to $\overleftrightarrow{QP}$? _ _ _ _ _

   Are lines $\overleftrightarrow{QP}$ and $\overleftrightarrow{MN}$ parallel? _ _ _ _ _

4. Construct a perpendicular to $\overleftrightarrow{MN}$ through point N.

   Label as L the point where the perpendicular intersects line $\overleftrightarrow{QP}$.

5. Compare segment $\overline{KM}$ and segment $\overline{LN}$.

   $\overline{KM}$ is _ _ _ _ _ _ _ _ _ _ _ _ _ _ _ _ _ $\overline{LN}$.

   (a) shorter than

   (b) congruent to

   (c) longer than

1. Check: Are the lines parallel? _ _ _ _ _

2. Construct the line parallel to $\overleftrightarrow{AB}$ through point Z.

Z

A

B

Corresponding Angles

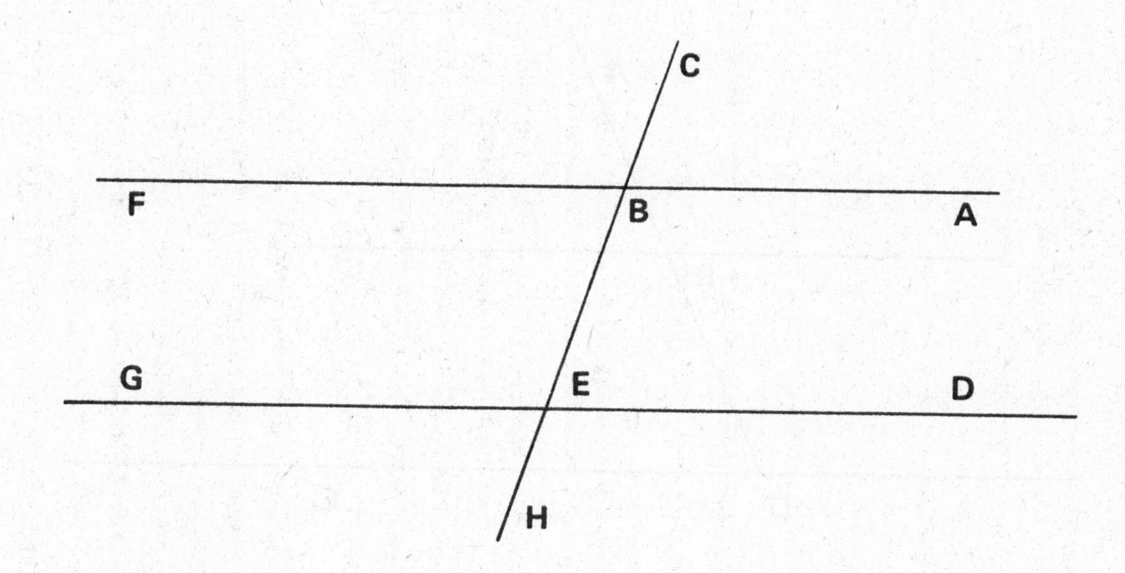

1. Check: Are lines $\overleftrightarrow{AF}$ and $\overleftrightarrow{DG}$ parallel? _ _ _ _ _

2. Angle CBA and angle BED are corresponding angles.

   Angle FBC and angle _ _ _ _ _ _ _ _ _ are corresponding angles.

   Angle FBE and angle _ _ _ _ _ _ _ _ _ are corresponding angles.

3. Compare angle CBA and angle BED.

   Angle CBA is _ _ _ _ _ _ _ _ _ _ _ _ _ _ _ _ angle BED.

   (a) smaller than (c) larger than

   (b) congruent to

4. Compare angle CBF and angle BEG.

   Are they congruent? _ _ _ _ _

5. If two lines are parallel, their corresponding angles are _ _ _ _ _ .

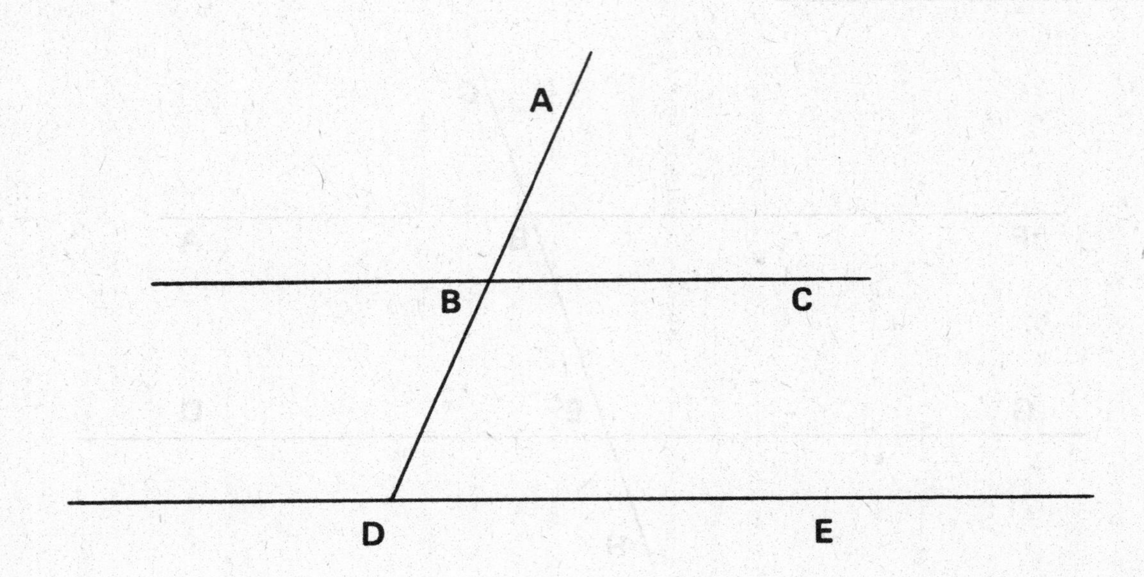

1. Compare angle ABC and angle BDE.

   Are they congruent? _ _ _ _ _

   Are they corresponding angles? _ _ _ _ _

2. Check: Is line $\overleftrightarrow{BC}$ parallel to line $\overleftrightarrow{DE}$? _ _ _ _ _

3. Draw a line intersecting both lines.

4. Compare a pair of corresponding angles.

   Are they congruent? _ _ _ _ _

5. Are the lines parallel? _ _ _ _ _

6. Two lines are _ _ _ _ _ _ _ _ _ _ _ _ _ _ _ _ _ parallel if their corresponding angles are congruent.

   (a) always (b) sometimes (c) never

Review

1. Construct an angle with vertex P and ray $\overrightarrow{PQ}$ as one side which is congruent to angle ABC.

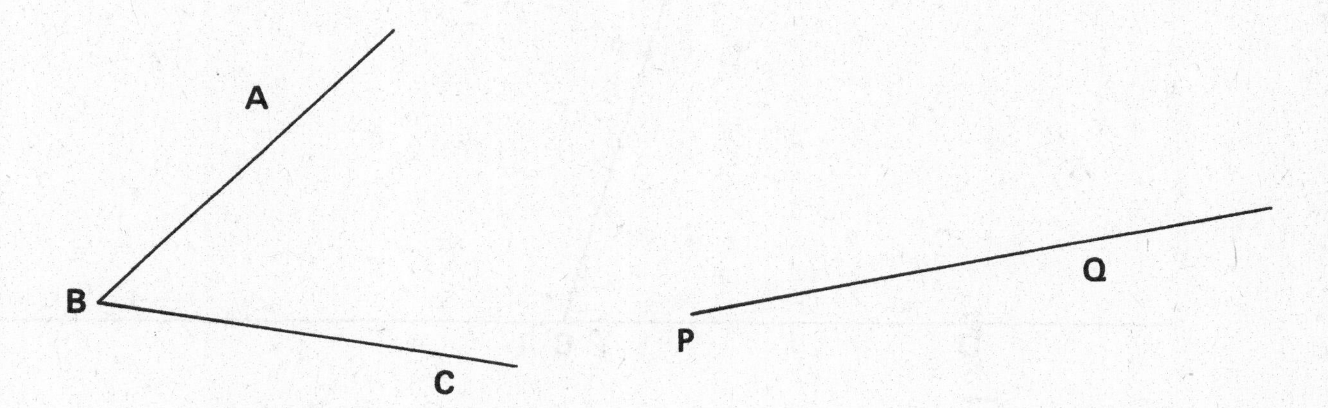

2. Construct a perpendicular to the given line through the given point.

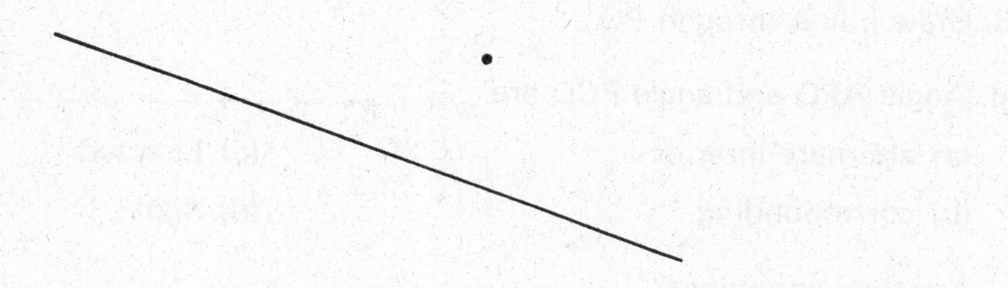

3. Construct a parallel to the given line through the given point.

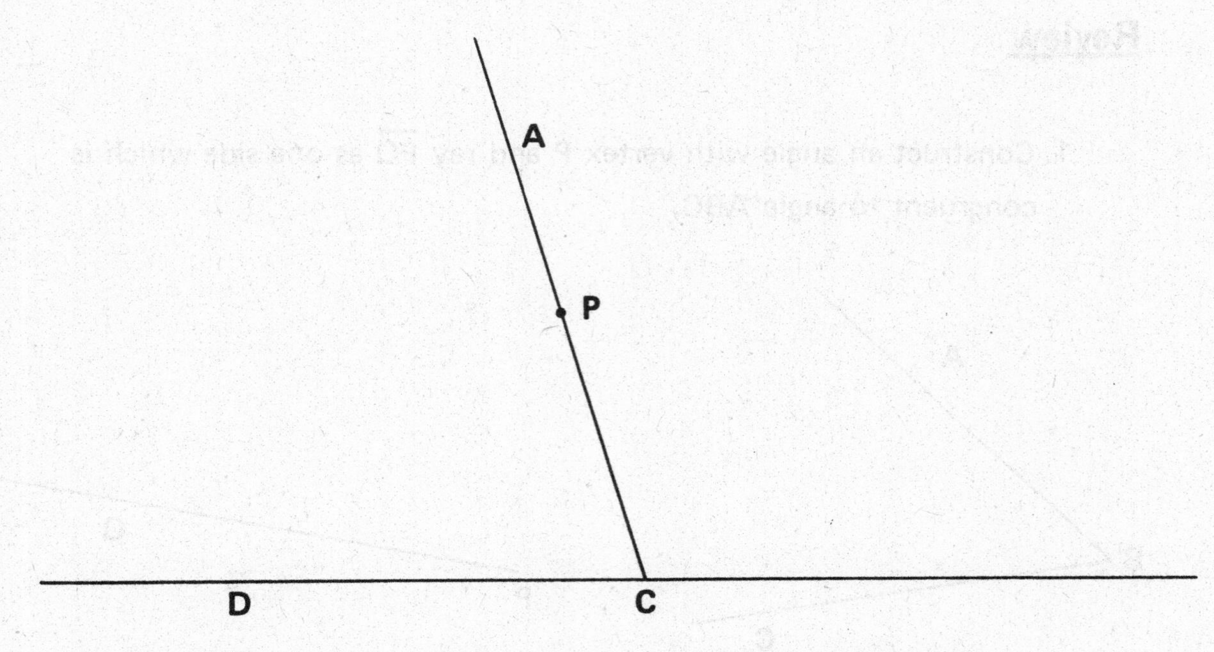

1. With ray $\overrightarrow{PA}$ as the right side, construct an angle at P which is congruent to angle PCD.

2. Label as Q a point on the angle to the left of P.

3. Draw a line through $\overline{PQ}$.

4. Angle APQ and angle PCD are _ _ _ _ _ _ _ _ _ _ _ _ _ angles.

   (a) alternate interior
   (b) corresponding
   (c) bisected
   (d) right

   Are they congruent? _ _ _ _ _

5. Check: Is line $\overleftrightarrow{PQ}$ parallel to line $\overleftrightarrow{CD}$? _ _ _ _ _

6. Construct another pair of congruent corresponding angles below.

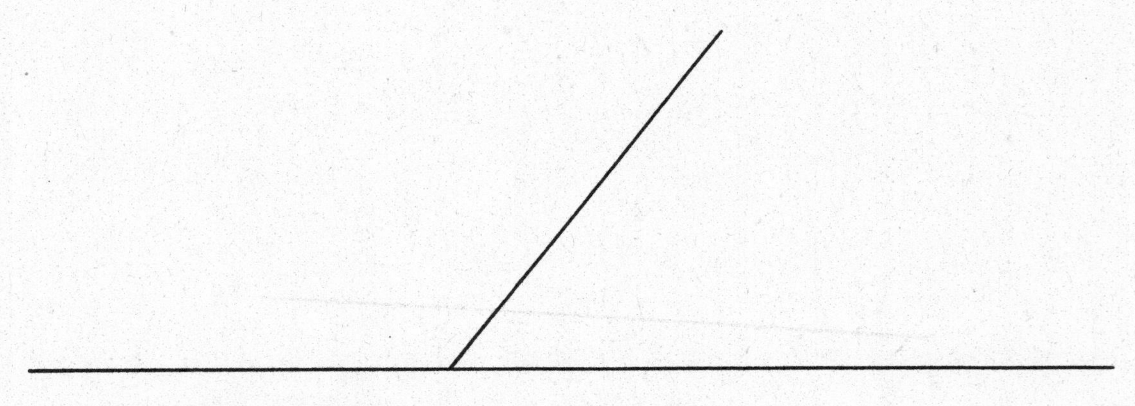

7. Label as $\overleftrightarrow{WX}$ and $\overleftrightarrow{YZ}$ a pair of parallel lines.

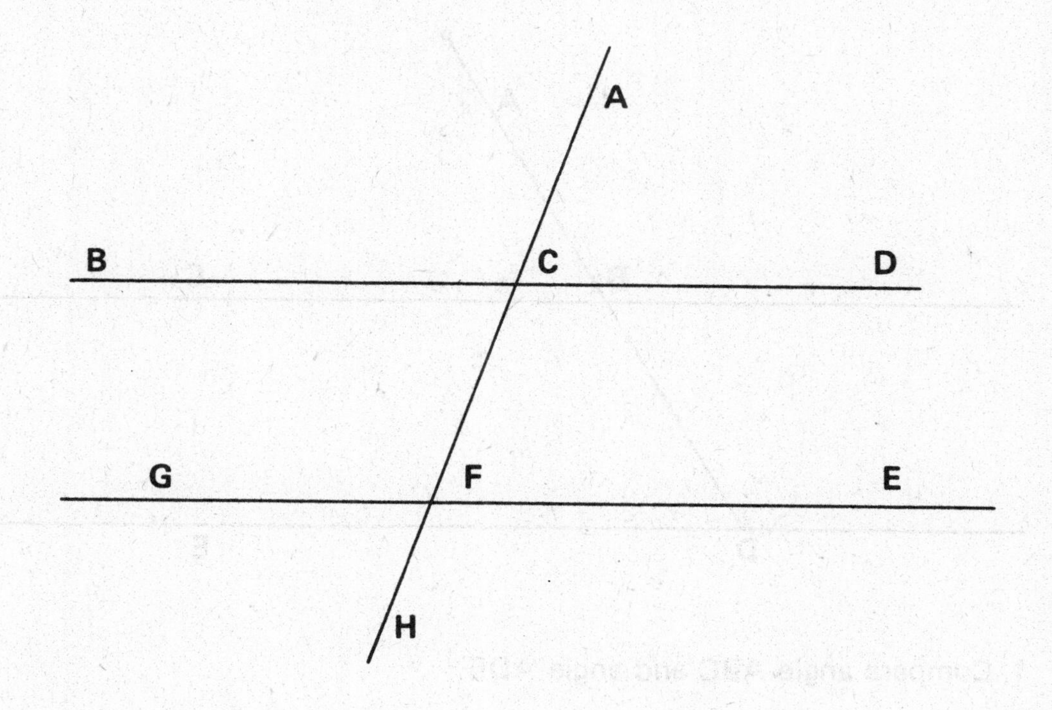

1. Compare angle ACD and angle CFE.

   Are they congruent? _ _ _ _ _

2. Are lines $\overleftrightarrow{BD}$ and $\overleftrightarrow{GE}$ parallel? _ _ _ _ _

   Why? _

3. Do you think angle GFH and angle CFE are congruent? _ _ _ _ _

   Why? _

4. Name some other angles congruent to angle CFE.

   _

5. Name three angles congruent to angle FCD.

   _

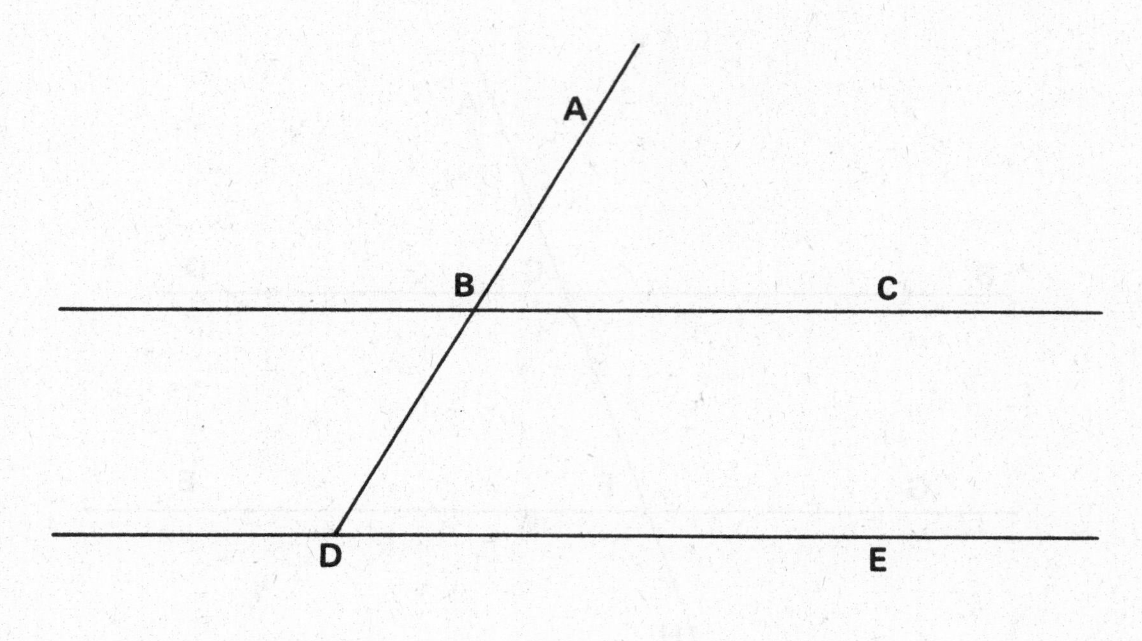

1. Compare angle ABC and angle ADE.

   Are they congruent? _ _ _ _ _

2. Are the lines $\overleftrightarrow{BC}$ and $\overleftrightarrow{DE}$ parallel? _ _ _ _ _

   Why? _

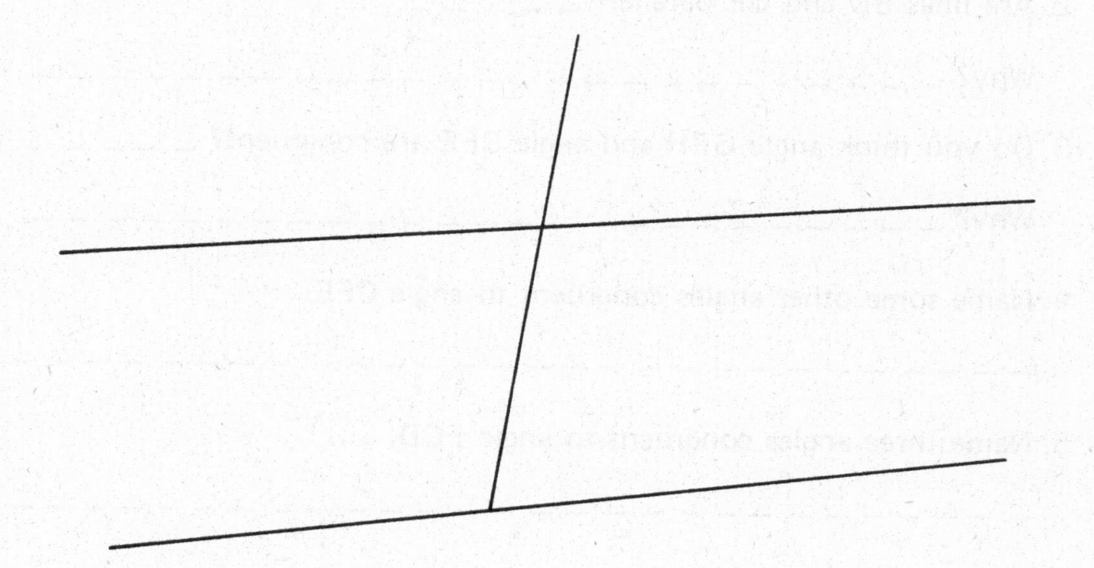

3. Check: Are the lines parallel? _ _ _ _ _

Alternate Interior Angles

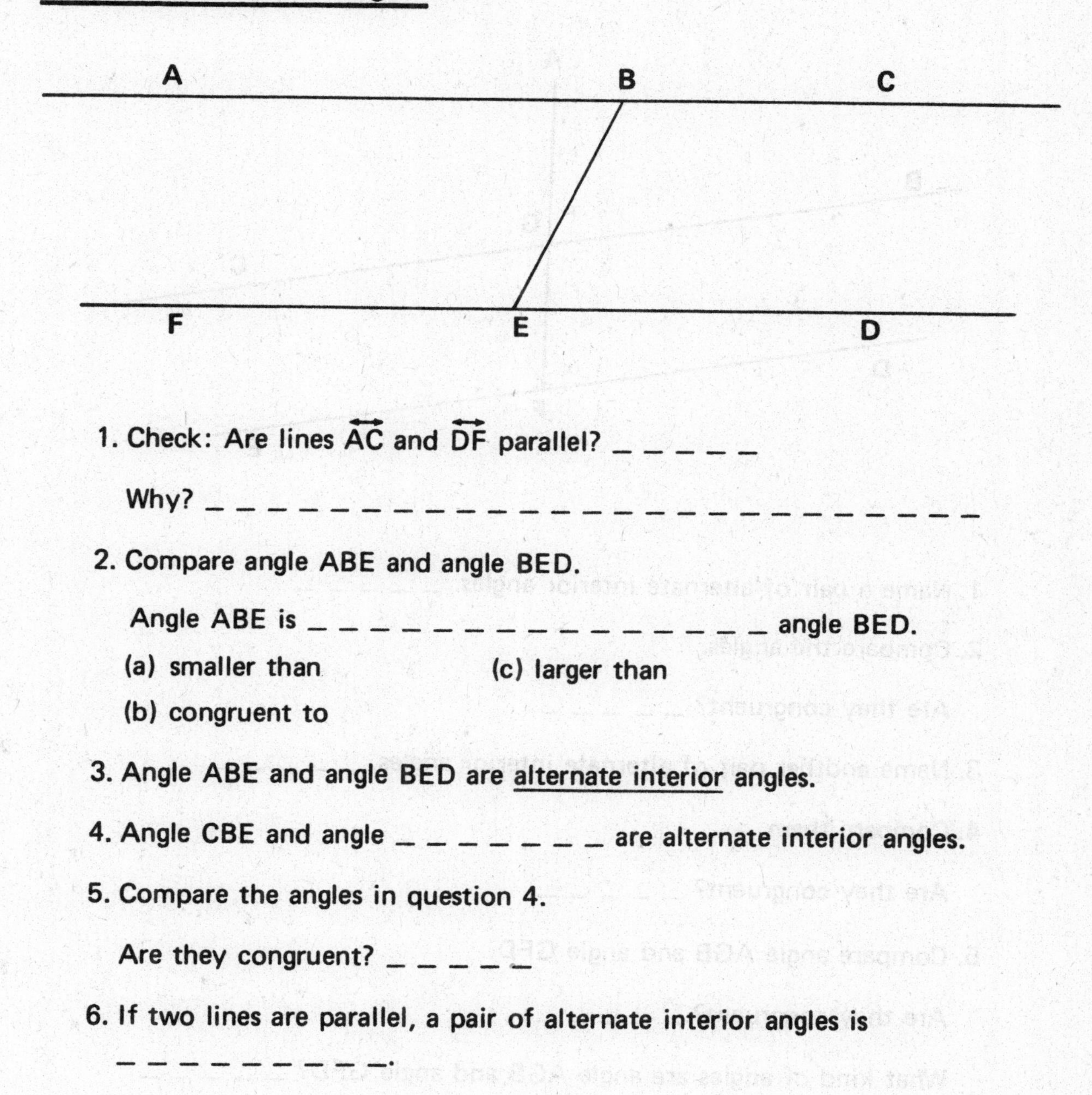

1. Check: Are lines $\overleftrightarrow{AC}$ and $\overleftrightarrow{DF}$ parallel? _ _ _ _ _ _

   Why? _

2. Compare angle ABE and angle BED.

   Angle ABE is _ _ _ _ _ _ _ _ _ _ _ _ _ _ _ _ _ _ _ angle BED.

   (a) smaller than (c) larger than

   (b) congruent to

3. Angle ABE and angle BED are <u>alternate interior</u> angles.

4. Angle CBE and angle _ _ _ _ _ _ _ _ _ are alternate interior angles.

5. Compare the angles in question 4.

   Are they congruent? _ _ _ _ _ _

6. If two lines are parallel, a pair of alternate interior angles is

   _ _ _ _ _ _ _ _ _ _ _.

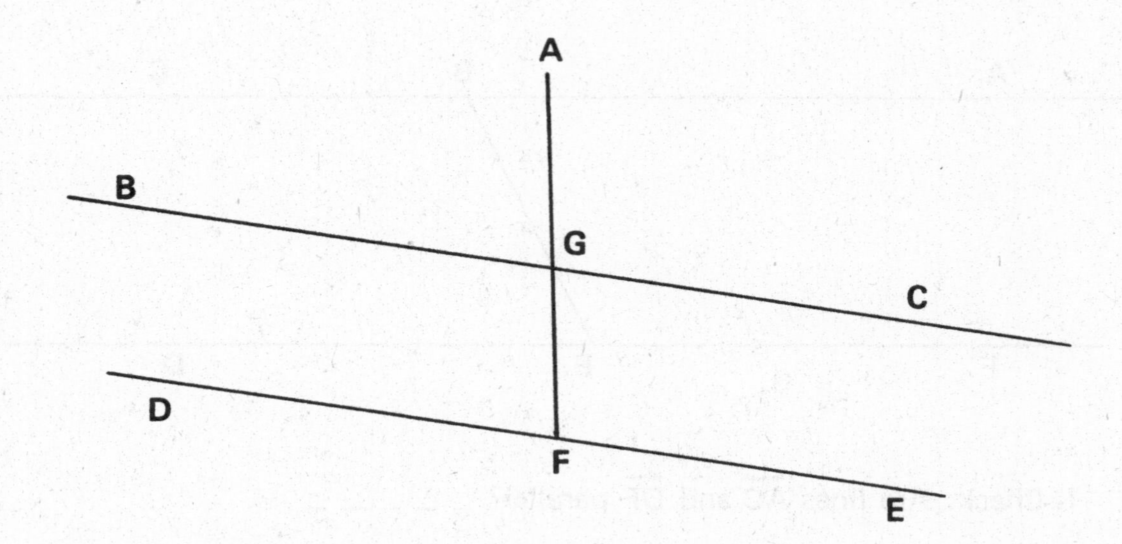

1. Name a pair of alternate interior angles. _ _ _ _ _

2. Compare the angles.

   Are they congruent? _ _ _ _ _

3. Name another pair of alternate interior angles. _ _ _ _ _

4. Compare them.

   Are they congruent? _ _ _ _ _

5. Compare angle AGB and angle GFD.

   Are they congruent? _ _ _ _ _

   What kind of angles are angle AGB and angle GFD? _ _ _ _ _

6. Are lines $\overleftrightarrow{BC}$ and $\overleftrightarrow{DE}$ parallel? _ _ _ _ _

   Why? _

Review

1. Construct the perpendicular bisector of the segment.

2. Construct the perpendicular to the line through the point.

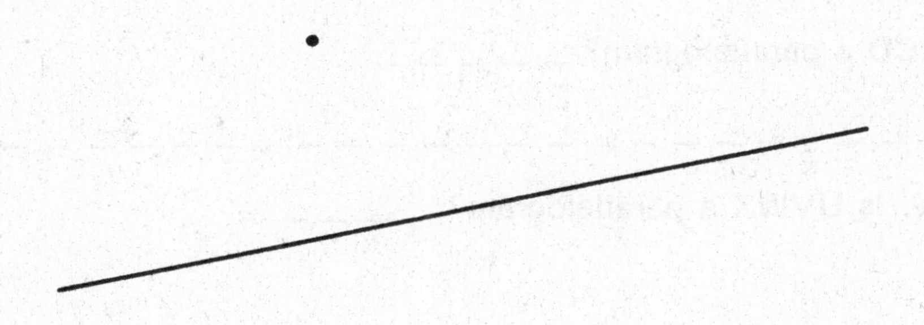

3. Check: Are the lines parallel? _____

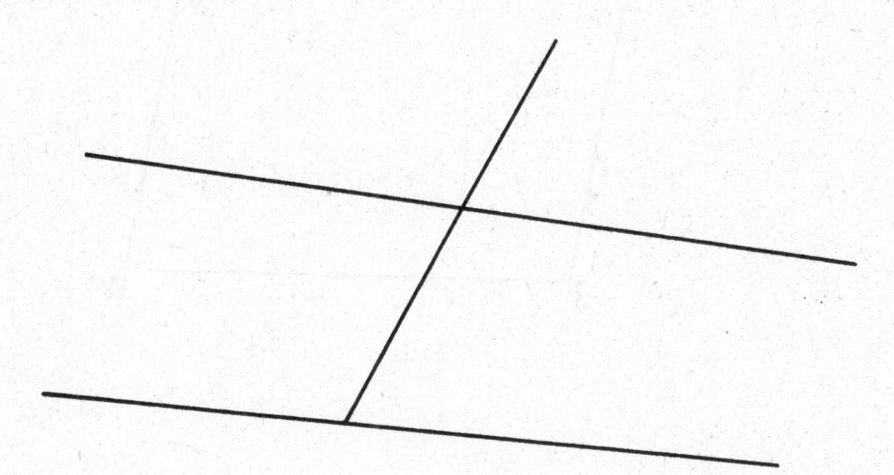

The Parallelogram

1. Check: Are $\overleftrightarrow{AB}$ and $\overleftrightarrow{DC}$ parallel? _ _ _ _ _

2. Check: Are $\overleftrightarrow{AD}$ end $\overleftrightarrow{BC}$ parallel? _ _ _ _ _

3. A parallelogram is a quadrilateral with two pairs of parallel sides.

   Is ABCD a parallelogram? _ _ _ _ _

   Why? _

4. Check: Is UVWX a parallelogram? _ _ _ _ _

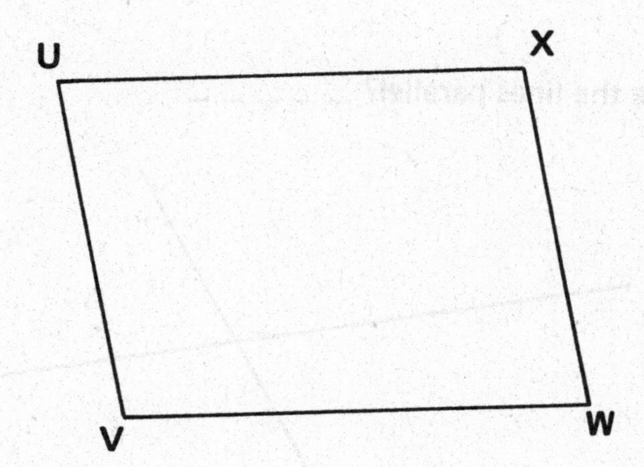

Which figures are parallelograms? _________________

____

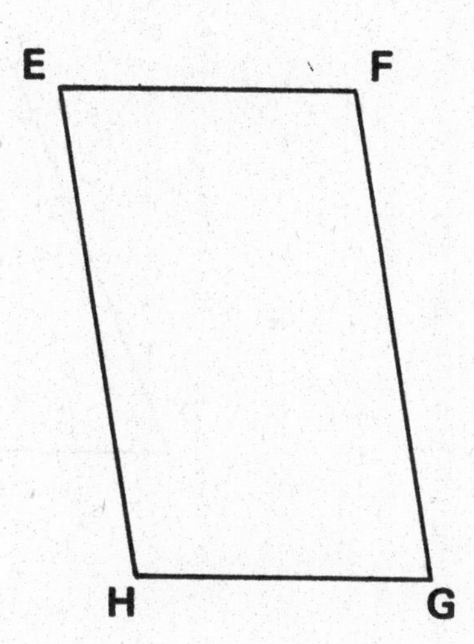

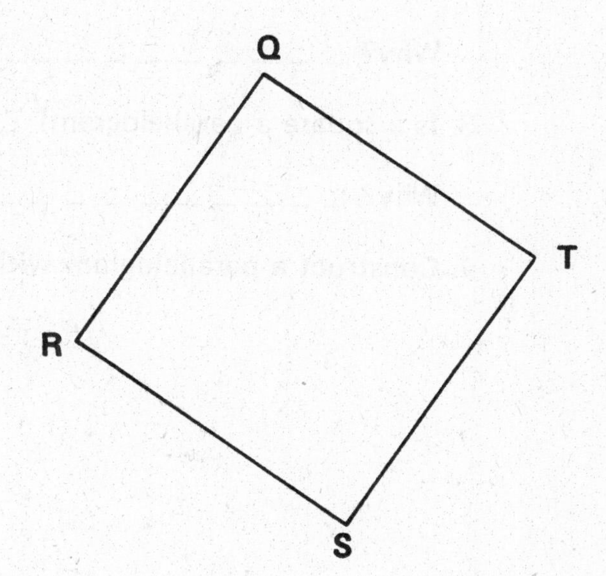

Constructing a Parallelogram

1. Construct a parallelogram with the two given sides:

2. Is a rectangle a parallelogram? _ _ _ _ _

   Why? _

3. Is a square a parallelogram? _ _ _ _ _

   Why? _

4. Construct a parallelogram with $\overline{AB}$ and $\overline{BC}$ as two of its sides.

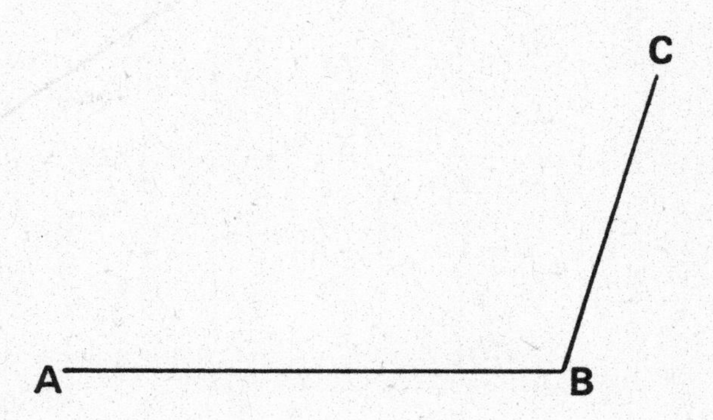

1. Construct a line parallel to $\overleftrightarrow{EF}$ through point X.

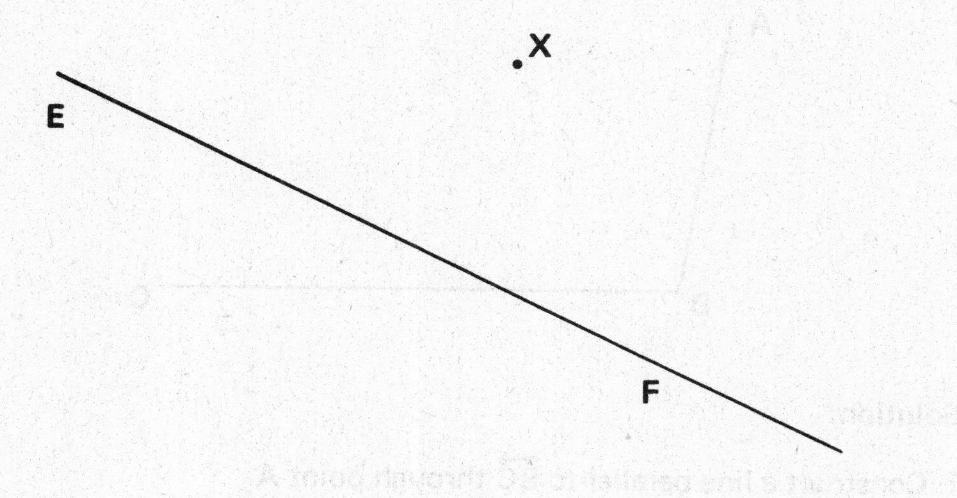

2. Check: Is ABCD a parallelogram? _ _ _ _ _

Why? _

Problem: *Construct a parallelogram with sides $\overline{AB}$ and $\overline{BC}$.*

Solution:

1. Construct a line parallel to $\overleftrightarrow{BC}$ through point A.

2. Construct a line parallel to $\overleftrightarrow{AB}$ through point C.

3. Label as D the point where the lines intersect.

4. Is ABCD a parallelogram? _ _ _ _ _

   Why? _

5. Construct a parallelogram with $\overline{WX}$ and $\overline{XY}$ as two sides.

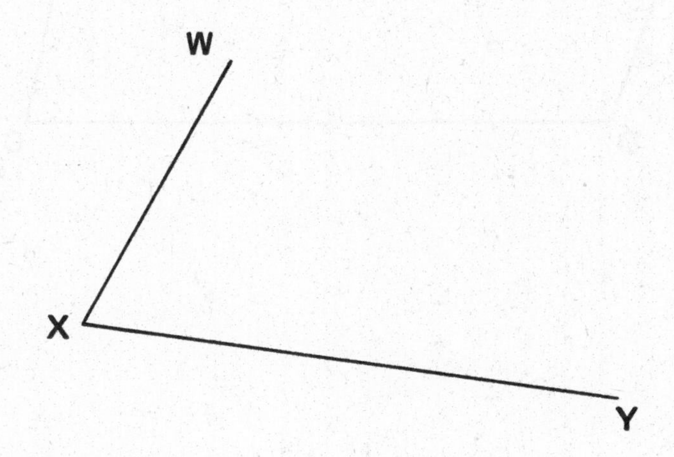

P •

A ——————————————— B

1. Through P construct a line parallel to $\overleftrightarrow{AB}$.
2. On the parallel to the right of P, lay off a segment with endpoint P which is congruent to $\overline{AB}$.
3. Label its other endpoint Q.
4. Draw $\overline{PA}$ and $\overline{QB}$.
5. Check: Is line $\overleftrightarrow{AB}$ parallel to line $\overleftrightarrow{PQ}$? _ _ _ _ _
6. Is $\overline{AB}$ congruent to $\overline{PQ}$? _ _ _ _ _
7. Is APQB a parallelogram? _ _ _ _ _

   Why? _

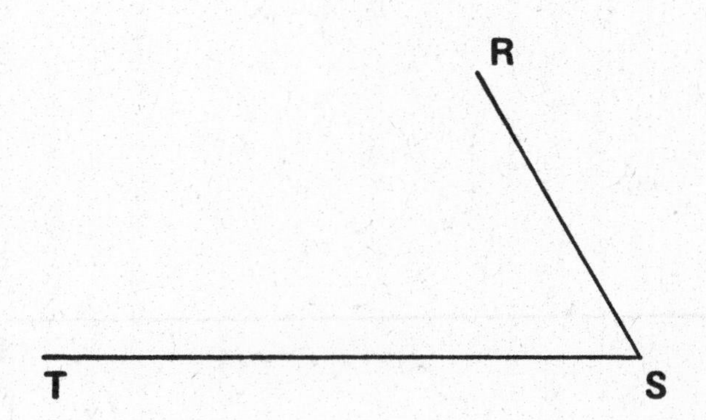

1. Draw an arc with center R and radius $\overline{ST}$ to the left of R.
2. Draw an arc with center T and radius $\overline{RS}$.
   Make the arcs intersect.
3. Label the intersection Q.
4. Draw $\overline{QR}$ and $\overline{QT}$.
5. Check: Is QRST a parallelogram? _ _ _ _ _

1. Construct a parallelogram with the two given sides.

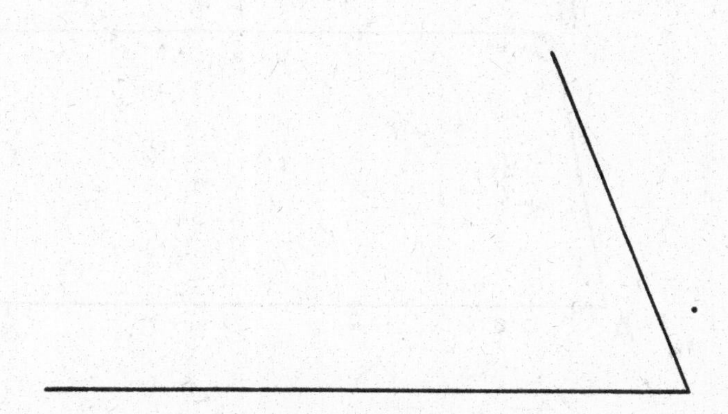

2. Use a different way to construct a parallelogram with the two given sides.

Properties of a Parallelogram

BARD is a parallelogram.

1. Compare sides $\overline{AR}$ and $\overline{BD}$.

   Are they congruent? _ _ _ _ _

2. Compare sides $\overline{BA}$ and $\overline{DR}$.

   Are they congruent? _ _ _ _ _

3. TOES is a parallelogram.

   Compare $\overline{TO}$ and $\overline{ES}$.

   Are they congruent? _ _ _ _ _

4. Compare $\overline{TS}$ and $\overline{OE}$.

   Are they congruent? _ _ _ _ _

5. The opposite sides of a parallelogram are _ _ _ _ _ _ _ _ _ _ _ _ _ congruent.

   (a) always (b) sometimes (c) never

ABCD is a parallelogram.

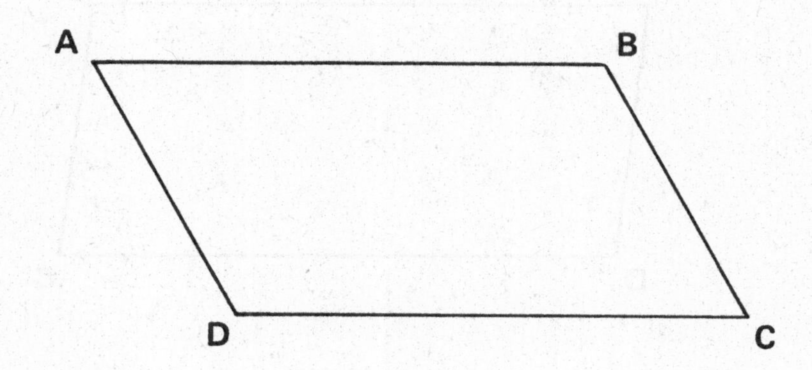

1. Compare angle BAD and angle BCD.

   Angle BAD is _ _ _ _ _ _ _ _ _ _ _ _ _ _ _ _ _ _ angle BCD.

   (a) smaller than
   (b) congruent to
   (c) larger than

2. Compare angle ABC and angle ADC.

   Angle ABC is _ _ _ _ _ _ _ _ _ _ _ _ _ _ _ _ _ _ angle ADC.

   (a) smaller than
   (b) congruent to
   (c) larger than

3. The opposite angles of a parallelogram are _ _ _ _ _ _ _ _ _ _ _ _ _.

   (a) congruent
   (b) parallel
   (c) right angles
   (d) vertical angles

4. Side $\overline{AB}$ is congruent to side _ _ _ _ _ _ .

   Side $\overline{AD}$ is congruent to side _ _ _ _ _ _ .

1. Check: Is ABCD a parallelogram? _ _ _ _ _

2. Draw diagonal $\overline{BD}$.

3. Trace triangle ABD.

   Is triangle ABD congruent to triangle BCD? _ _ _ _ _

4. Draw diagonal $\overline{AC}$.
   Label the point of intersection of the diagonals as M.

5. Trace triangle ABM.

   Triangle ABM is congruent to triangle _ _ _ _ _.

1. ABCD is a rectangle.

Check: Is ABCD a parallelogram? _ _ _ _ _

2. Compare $\overline{AB}$ and $\overline{CD}$.

Are they congruent? _ _ _ _ _

3. Compare $\overline{AD}$ and $\overline{BC}$.

Are they congruent? _ _ _ _ _

4. Compare angle BAD and angle BCD.

Are they congruent? _ _ _ _ _

5. Compare angle ADC and angle ABC.

Are they congruent? _ _ _ _ _

6. The opposite sides of a rectangle are _ _ _ _ _ _ _ _ _ _ _ _ _ _ parallel.

(a) always (b) sometimes (c) never

7. The opposite sides of a rectangle are _ _ _ _ _ _ _ _ _ _ _ _ _ _ congruent.

(a) always (b) sometimes (c) never

8. The opposite angles of a rectangle are _ _ _ _ _ _ _ _ _ _ _ _ _ congruent?

(a) always (b) sometimes (c) never

PARL is a parallelogram.

1. Draw diagonals $\overline{PR}$ and $\overline{AL}$.

2. Label their point of intersection B.

3. Compare $\overline{BP}$ and $\overline{BR}$.

   Is B the midpoint of $\overline{PR}$? _ _ _ _ _

4. Compare $\overline{AB}$ and $\overline{BL}$.

   Is B the midpoint of $\overline{AL}$? _ _ _ _ _

5. Compare $\overline{PR}$ and $\overline{AL}$.

   Are they congruent? _ _ _ _ _

6. $\overline{AL}$ _ _ _ _ _ _ _ _ _ _ _ _ _ _ _ _ _ _ _ $\overline{PR}$.

   (a) is congruent to

   (b) bisects

   (c) is the perpendicular bisector of

1. WXYZ is a rectangle.

   Is WXYZ a parallelogram? _ _ _ _ _

   Why? _

2. Draw diagonals $\overline{WY}$ and $\overline{ZX}$.

3. Compare $\overline{WY}$ and $\overline{ZX}$.

   Are the diagonals congruent? _ _ _ _ _

4. Check: Is MOPS a parallelogram? _ _ _ _ _

5. Draw diagonals $\overline{MP}$ and $\overline{OS}$.

6. Compare $\overline{MP}$ and $\overline{OS}$.

   Are they congruent? _ _ _ _ _

7. The diagonals of a parallelogram are _ _ _ _ _ _ _ _ _ _ _ _ _ _ _ congruent.

   (a) always (b) sometimes (c) never

The Rhombus

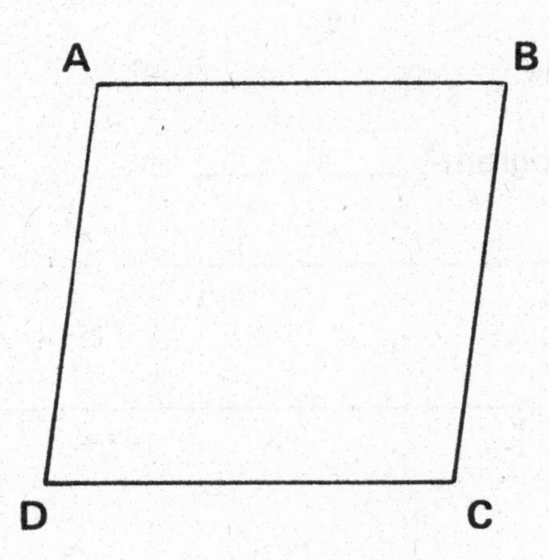

1. Compare $\overline{AB}$, $\overline{BC}$, $\overline{CD}$, and $\overline{AD}$.

   Are they all congruent? _ _ _ _ _

2. A rhombus is a parallelogram with four congruent sides.

   Which of the figures are rhombuses? _ _ _ _ _ _ _ _ _ _ _ _ _

   _

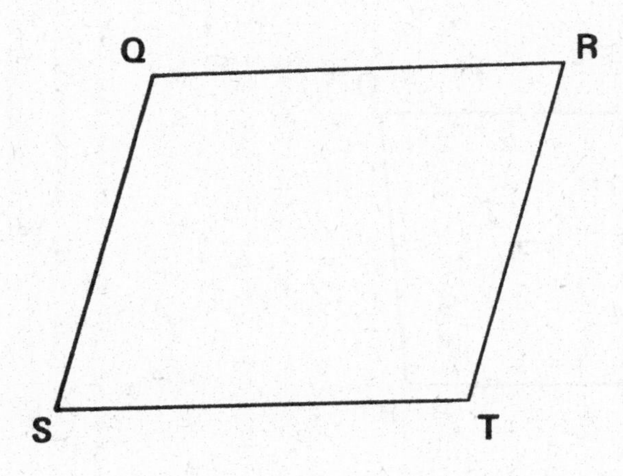

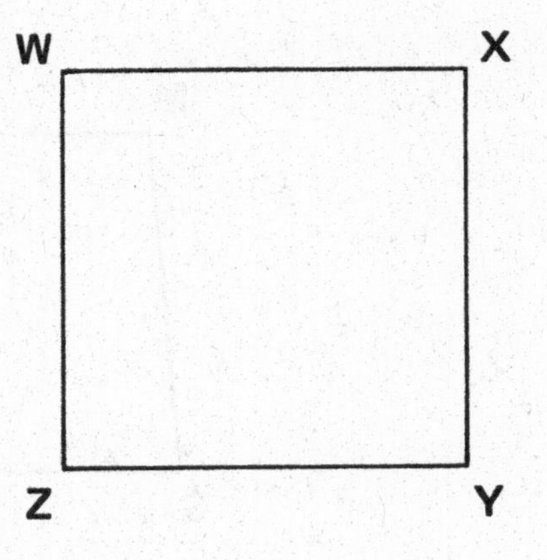

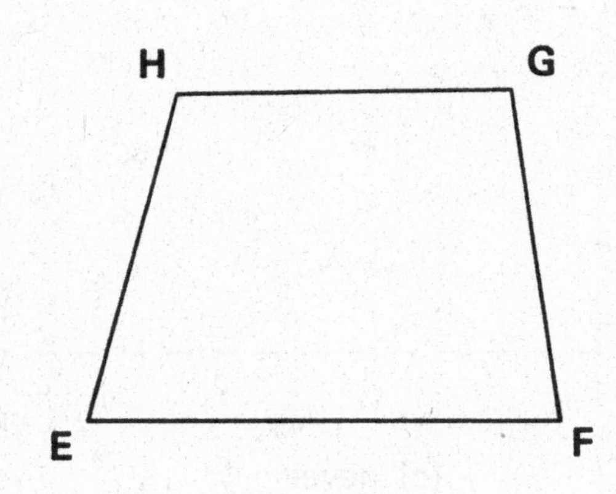

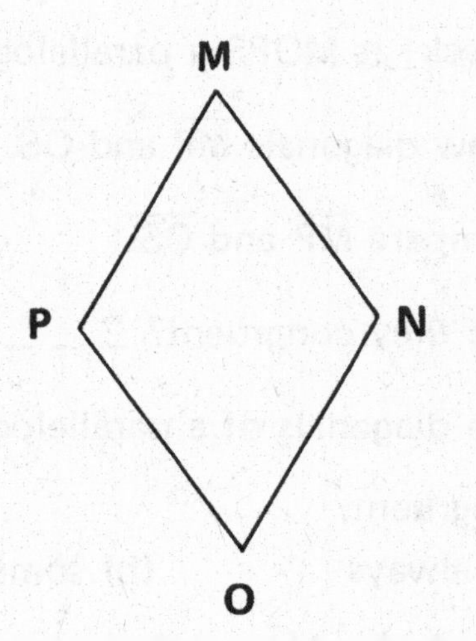

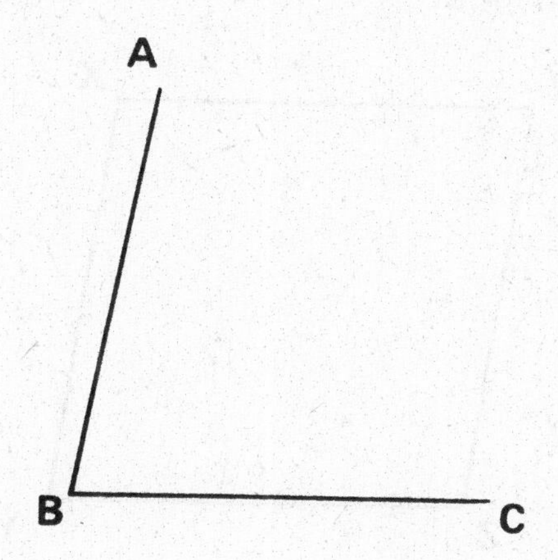

1. Check: Are segments $\overline{AB}$ and $\overline{BC}$ congruent? _ _ _ _ _

2. Construct the rest of the parallelogram with $\overline{AB}$ and $\overline{BC}$ as sides.

3. Label the fourth vertex D.

4. Is ABCD a rhombus? _ _ _ _ _

   Why? _

5. Construct another rhombus.

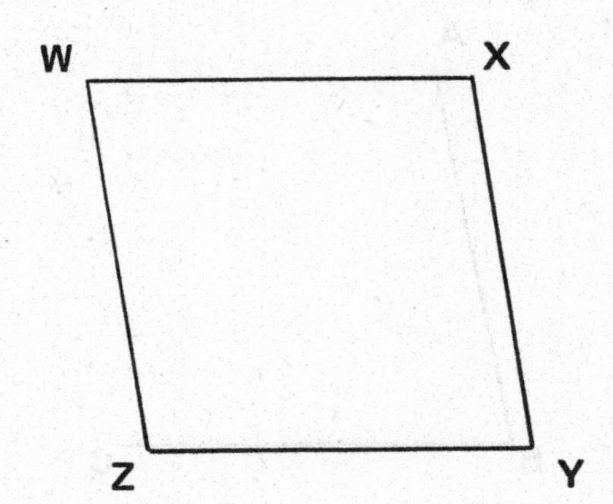

1. Check: Are all four sides of WXYZ congruent? _ _ _ _ _

2. Is WXYZ a rhombus? _ _ _ _ _

   Why? _

3. Draw diagonal $\overleftrightarrow{WY}$.

4. Check: Does $\overleftrightarrow{WY}$ bisect angle XYZ? _ _ _ _ _

5. Draw diagonal $\overline{XZ}$.

6. Check: Is $\overleftrightarrow{XZ}$ perpendicular to $\overleftrightarrow{WY}$? _ _ _ _ _

7. Check: Does $\overleftrightarrow{XZ}$ bisect segment $\overline{WY}$? _ _ _ _ _

   Check: Does $\overleftrightarrow{WY}$ bisect segment $\overline{XZ}$? _ _ _ _ _

8. The diagonals of a rhombus are _ _ _ _ _ _ _ _ _ _ _ _ _ _ _ _ _ _.

   (a) congruent

   (b) perpendicular bisectors of each other

   (c) parallel

Review

1. Find the midpoint of the segment.

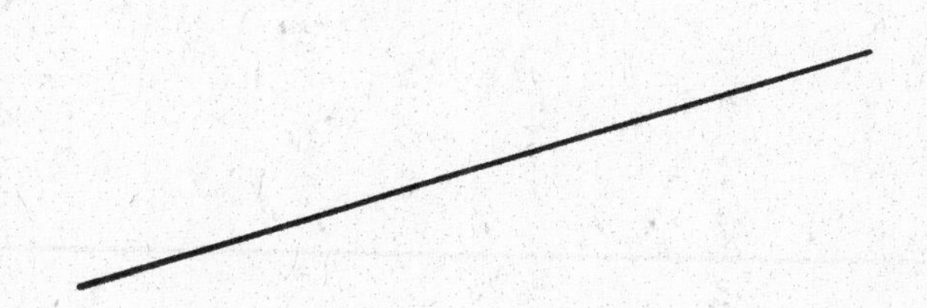

2. Construct a rectangle.

The Midpoints of a Quadrilateral

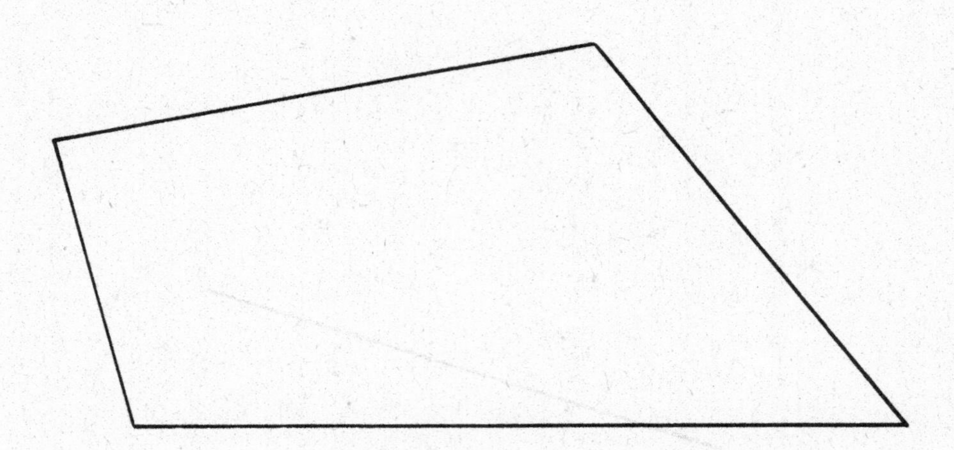

1. Bisect each side of the quadrilateral.

2. Label the midpoints M, N, O, P in that order.

3. Draw MNOP.

4. Compare the sides of MNOP.

   Side $\overline{MN}$ is congruent to _ _ _ _ _.

   Side $\overline{NO}$ is congruent to _ _ _ _ _.

5. What kind of figure is MNOP? _ _ _ _ _ _ _ _ _ _ _ _ _ _ _ _ _ _

   (a) triangle
   (b) square
   (c) rectangle
   (d) parallelogram

1. What kind of figure is ABCD? _ _ _ _ _ _ _ _ _ _ _ _ _ _ _ _ _

   (a) square
   (b) equilateral triangle
   (c) rectangle
   (d) rhombus

2. Bisect each side of ABCD.

3. Label the midpoints W, X, Y, Z in that order.

4. WXYZ is a _ _ _ _ _ _ _ _ _ _ _ _ _ _ _ _ _.

   (a) square
   (b) equilateral triangle
   (c) rectangle
   (d) rhombus

5. How do you know?_ _

Review

1. Construct a parallelogram with the two given sides.

2. Construct a rhombus.

3. Construct a line parallel to the given line through the given point.

The Sum of Angles

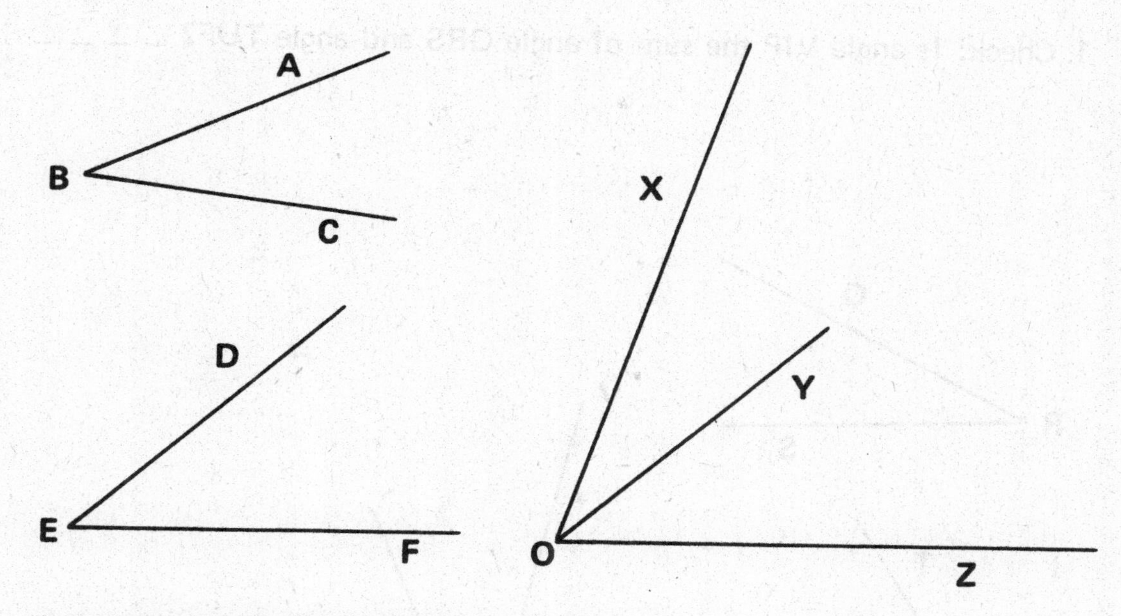

1. Compare angle XOY and angle ABC.

   Are they congruent? _ _ _ _ _

2. Compare angle YOZ and angle DEF.

   Are they congruent? _ _ _ _ _

3. Angle XOZ is the sum of angle ABC and angle DEF.

   Check: Is angle MPN the sum of angle UVW and angle XYZ? _ _ _ _

   Why? _

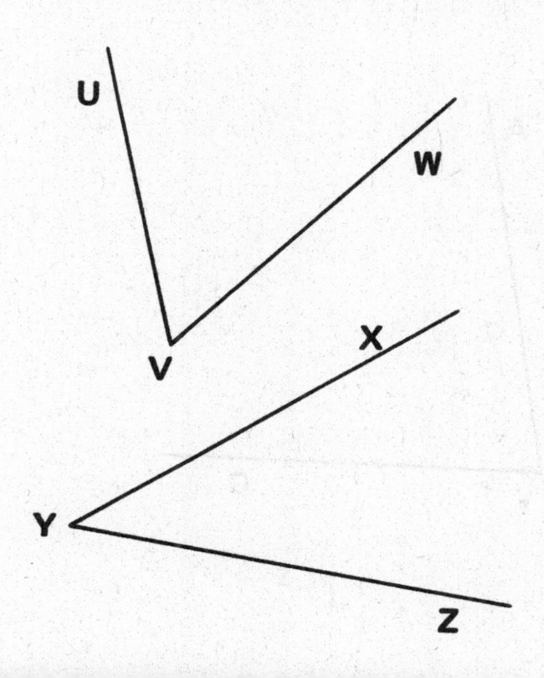

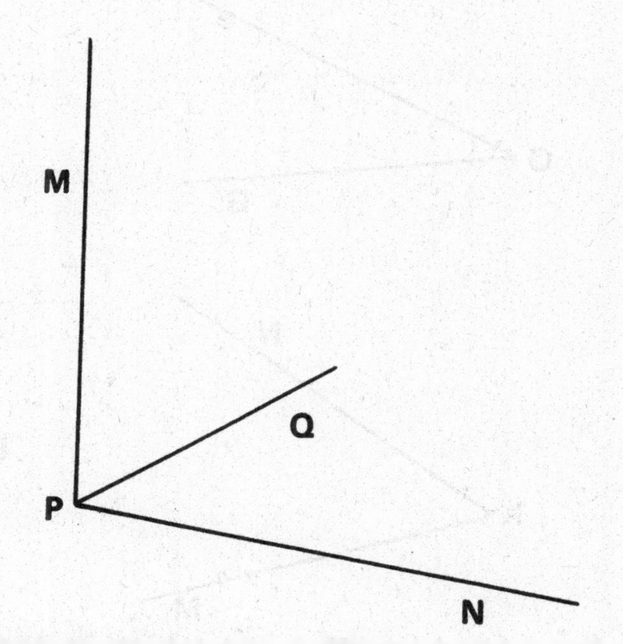

1. Check: Is angle VIP the sum of angle QRS and angle TUF? _ _ _ _

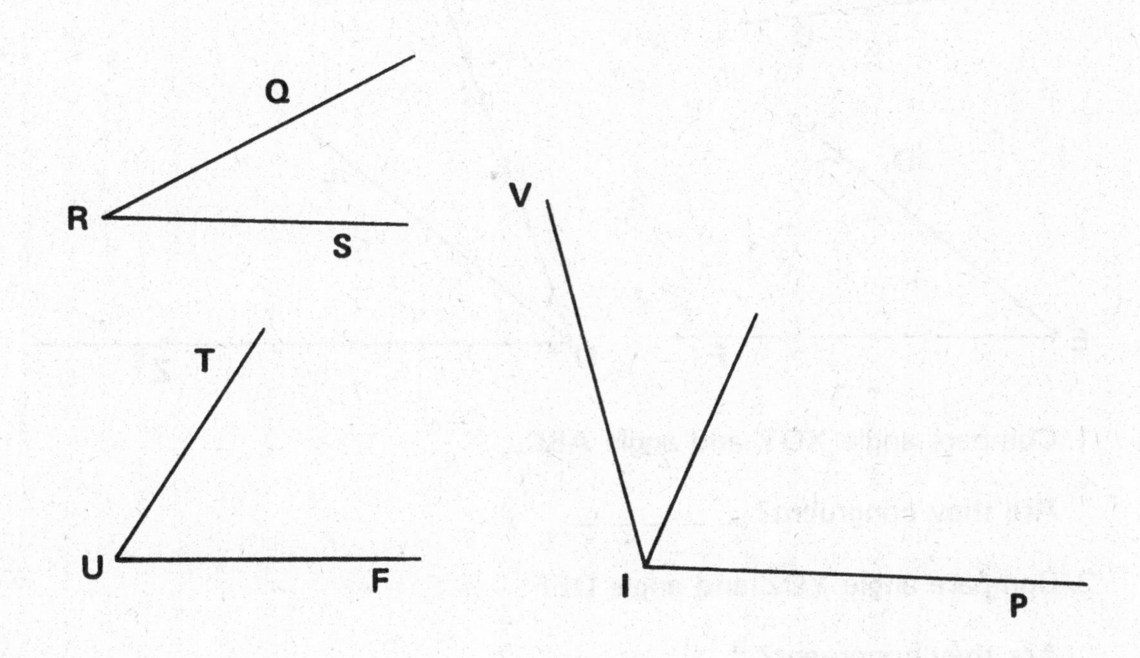

2. Check: Is angle ABC the sum of angle DOG and angle NKM? _ _ _ _

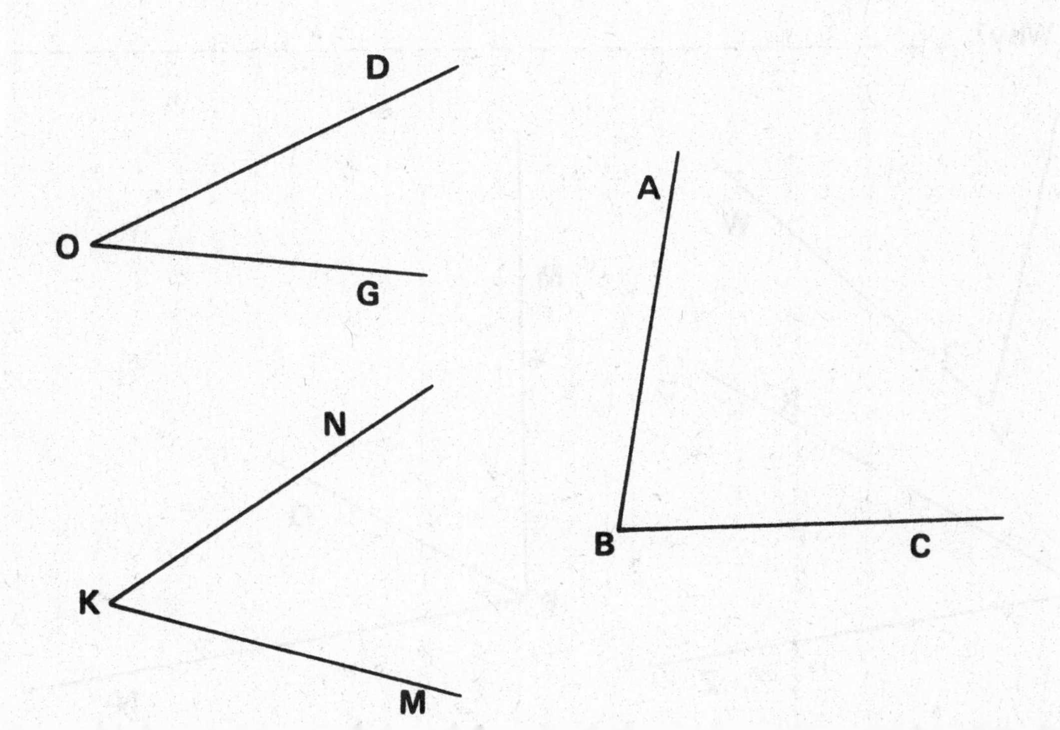

Constructing the Sum of Angles

Problem: *Find the sum of the angles.*

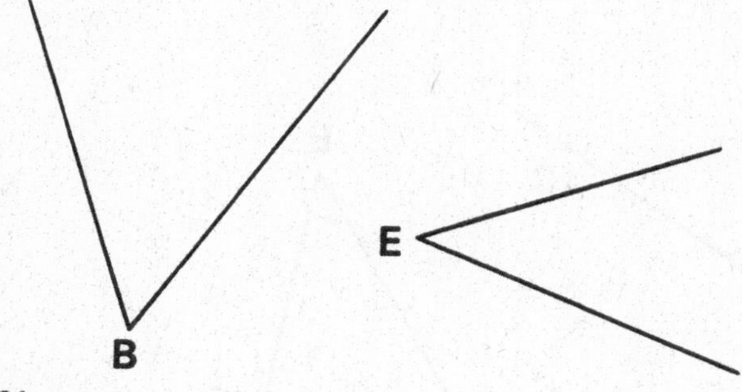

Solution:

P •

1. Draw a ray from P to the right.
2. Draw a large arc above the ray, and make it intersect the ray.
   Label as Q the point of intersection.
3. Use the same radius to draw an arc with center B intersecting both sides of the angle at B.
   Label the points of intersection A and C.
4. Use the same radius to draw an arc with center E intersecting both sides of the angle at E.
   Label the points of intersection D and F.
5. Draw an arc with center Q and radius $\overline{AC}$ intersecting the arc through Q above Q.
   Label the point of intersection R.
6. Draw an arc with center R and radius $\overline{DF}$ intersecting the arc through Q and R above R.
   Label the point of intersection T.
7. Draw $\overline{TP}$.
8. Now draw $\overline{RP}$.

Angle RPQ is congruent to angle _ _ _ _ _.

Angle TPR is congruent to angle _ _ _ _ _.

Angle TPQ is the sum of angle _ _ _ _ _ _ and angle _ _ _ _ _.

1. Construct the sum of angle ABC and angle DEF.

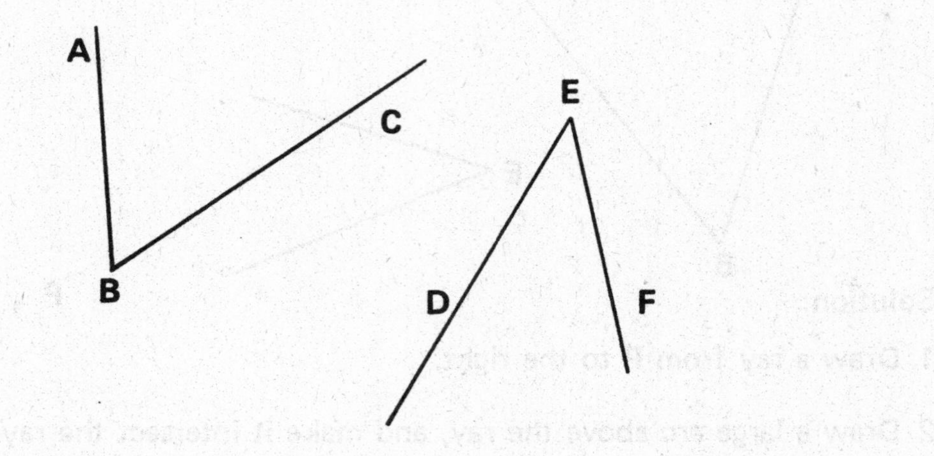

2. Construct the sum of the three angles.

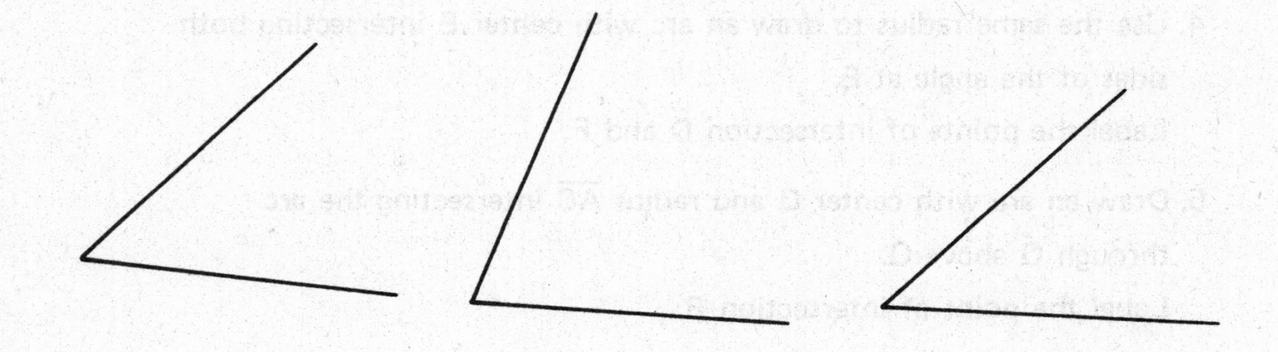

The Exterior Angles of a Triangle

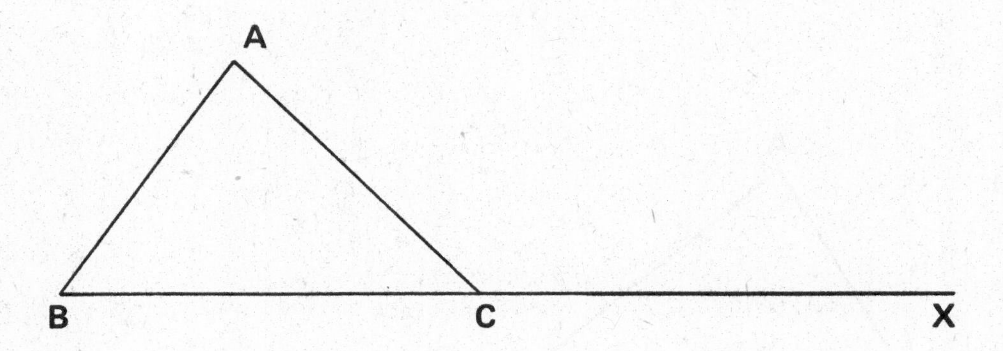

Angle ACX is an <u>exterior</u> angle of triangle ABC.

1. Extend side $\overline{AC}$ beyond A.

   Label as Y a point beyond a on ray $\overleftarrow{AC}$.

   Angle _ _ _ _ _ is another exterior angle of triangle ABC.

2. Extend side $\overline{AB}$ beyond B.

   Label as Z a point on ray $\overrightarrow{AB}$ beyond B.

   Name another exterior angle of triangle ABC. _ _ _ _ _

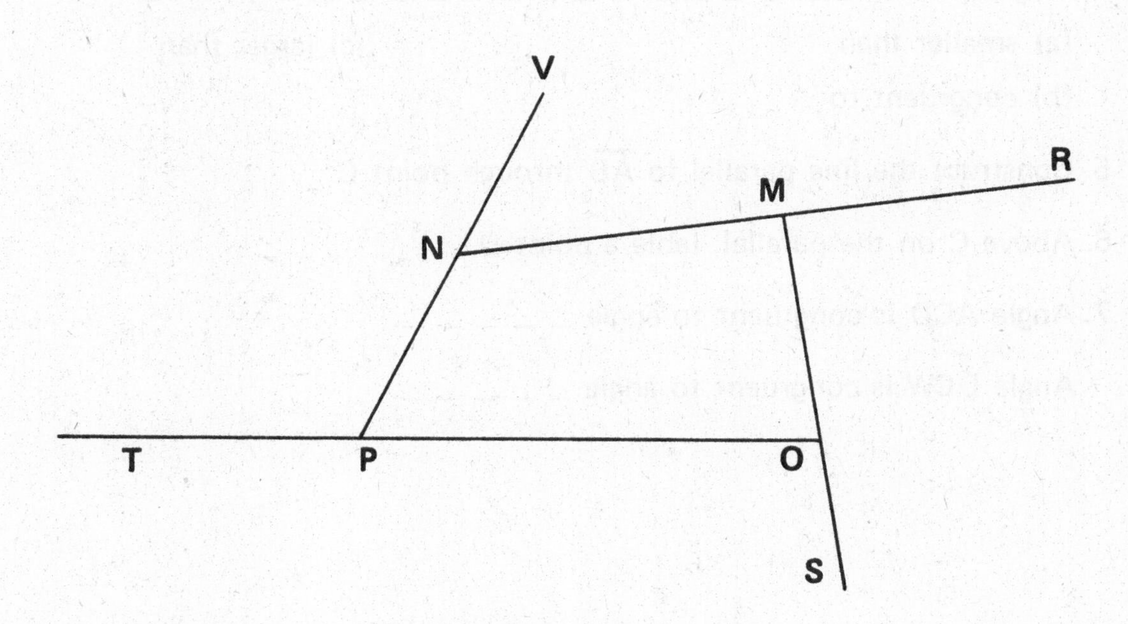

3. Name four exterior angles of quadrilateral NMOP.

   _

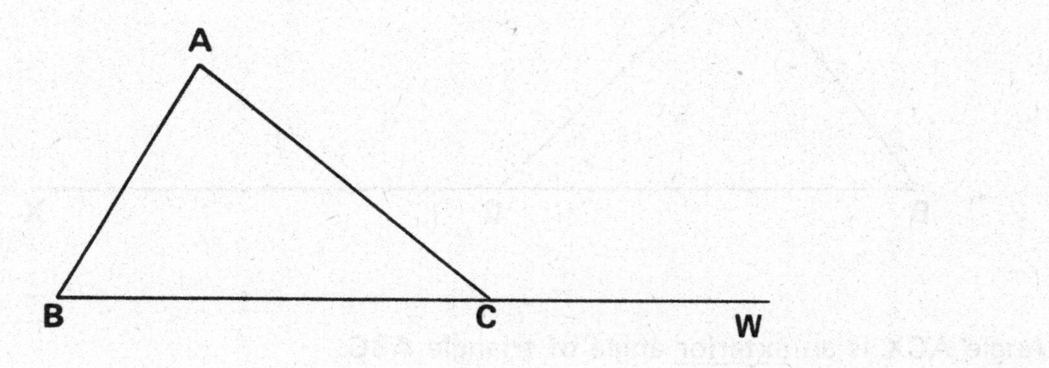

1. Name an exterior angle of triangle ABC. _ _ _ _ _

2. Find the sum of angle ABC and angle BAC.

3. Label their sum angle XYZ.

4. Compare angle ACW and angle XYZ.

   Angle ACW is _ _ _ _ _ _ _ _ _ _ _ _ _ _ _ _ _ angle XYZ.

   (a) smaller than
   (b) congruent to
   (c) larger than

5. Construct the line parallel to $\overleftrightarrow{AB}$ through point C.

6. Above C on the parallel, lable a point D.

7. Angle ACD is congruent to angle _ _ _ _ _.

   Angle DCW is congruent to angle _ _ _ _ _.

The Sum of the Exterior Angles of a Triangle

Problem: *Find the sum of the exterior angles of a triangle.*

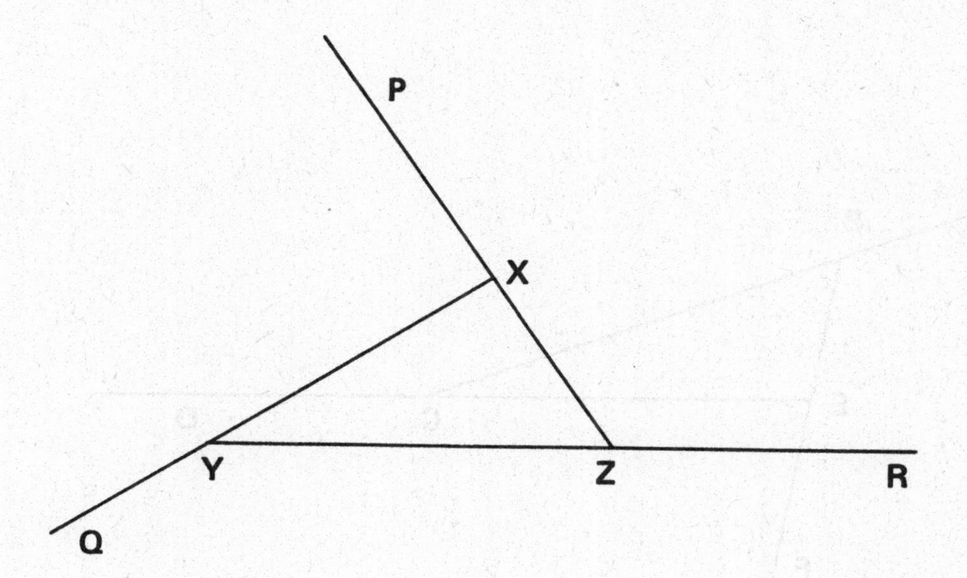

A B

Solution:

1. Reproduce angle XZR at A with $\overrightarrow{AB}$ as one side. Draw the other side above line $\overleftrightarrow{AB}$.
2. Label the reproduced angle CAB.
3. Reproduce angle PXY at A with $\overrightarrow{AC}$ as one side. Draw its other side to the left of line $\overleftrightarrow{AC}$.
4. Label the reproduced angle DAC.
5. Reproduce angle QYZ at A with $\overrightarrow{AD}$ as one side. Draw its other side to the right of $\overleftrightarrow{AD}$.
6. Does its other side lie on ray $\overrightarrow{AB}$? _ _ _ _ _

Construct the sum of the exterior angles BCD, CEF, and ABE.

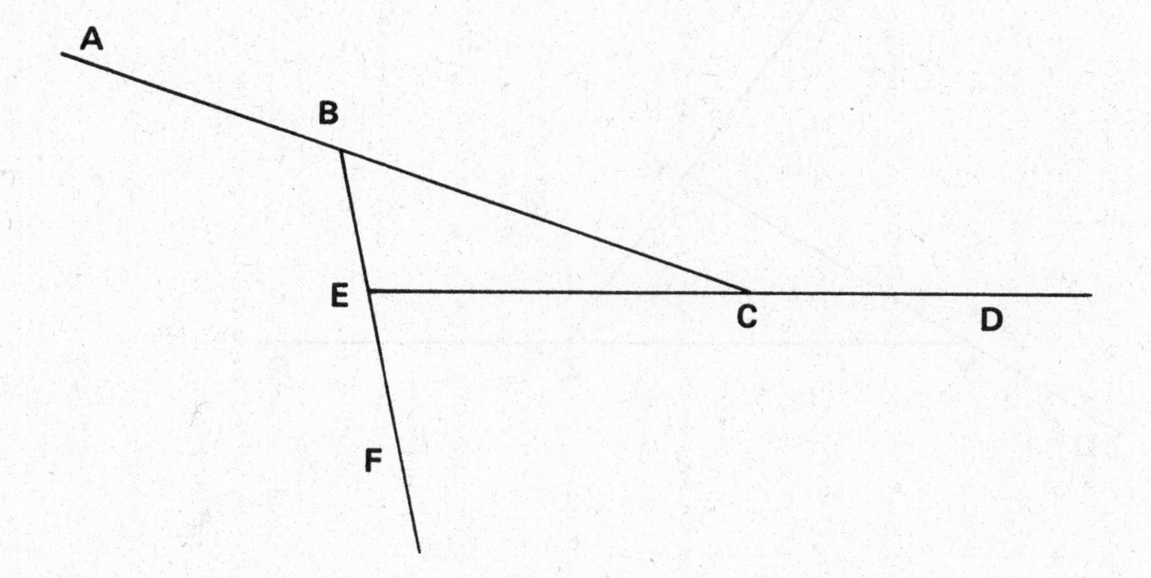

The Sum of the Exterior Angles of a Polygon

Construct the sum of the four exterior angles of quadrilateral ABCD.

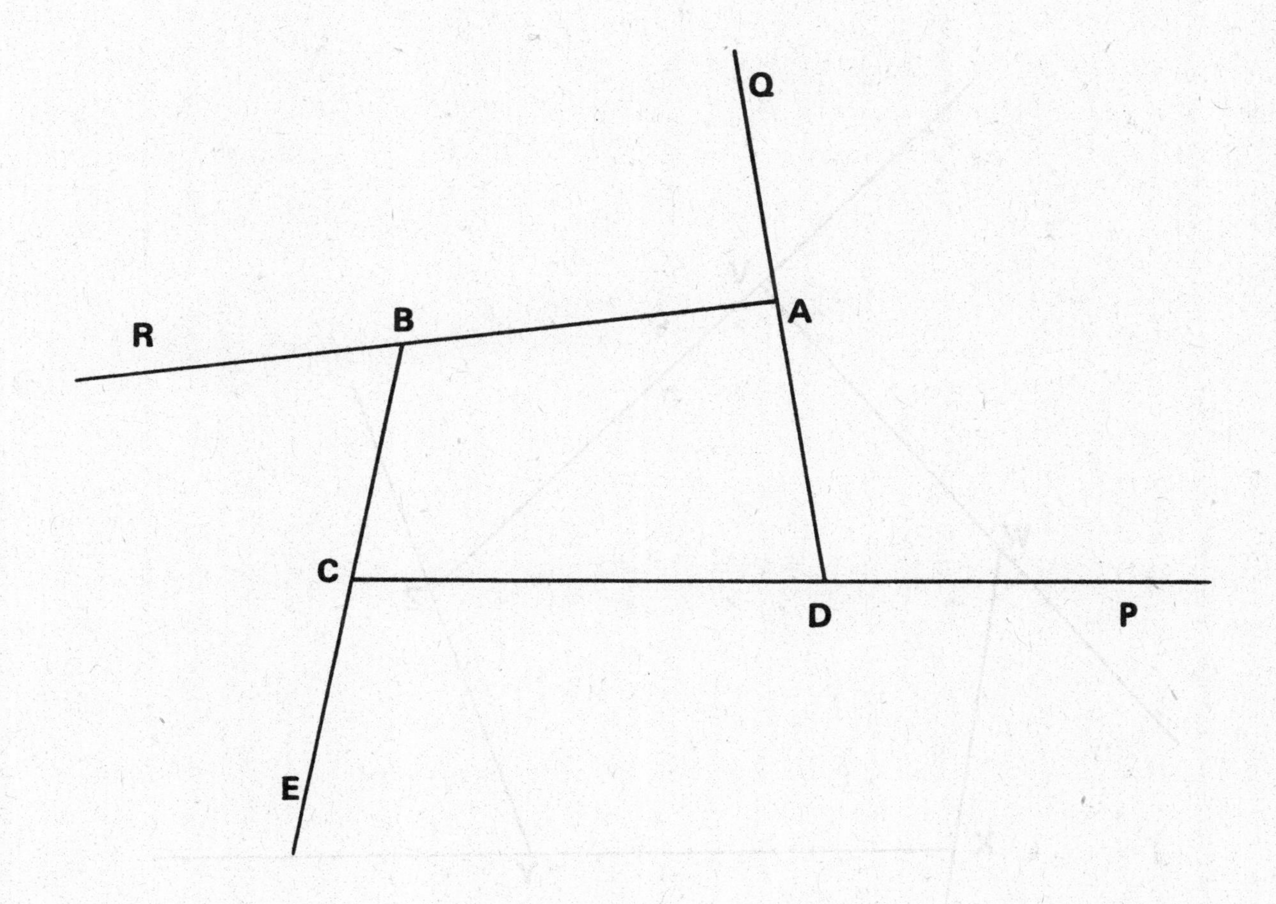

Construct the sum of the exterior angles of pentagon VWXYZ.

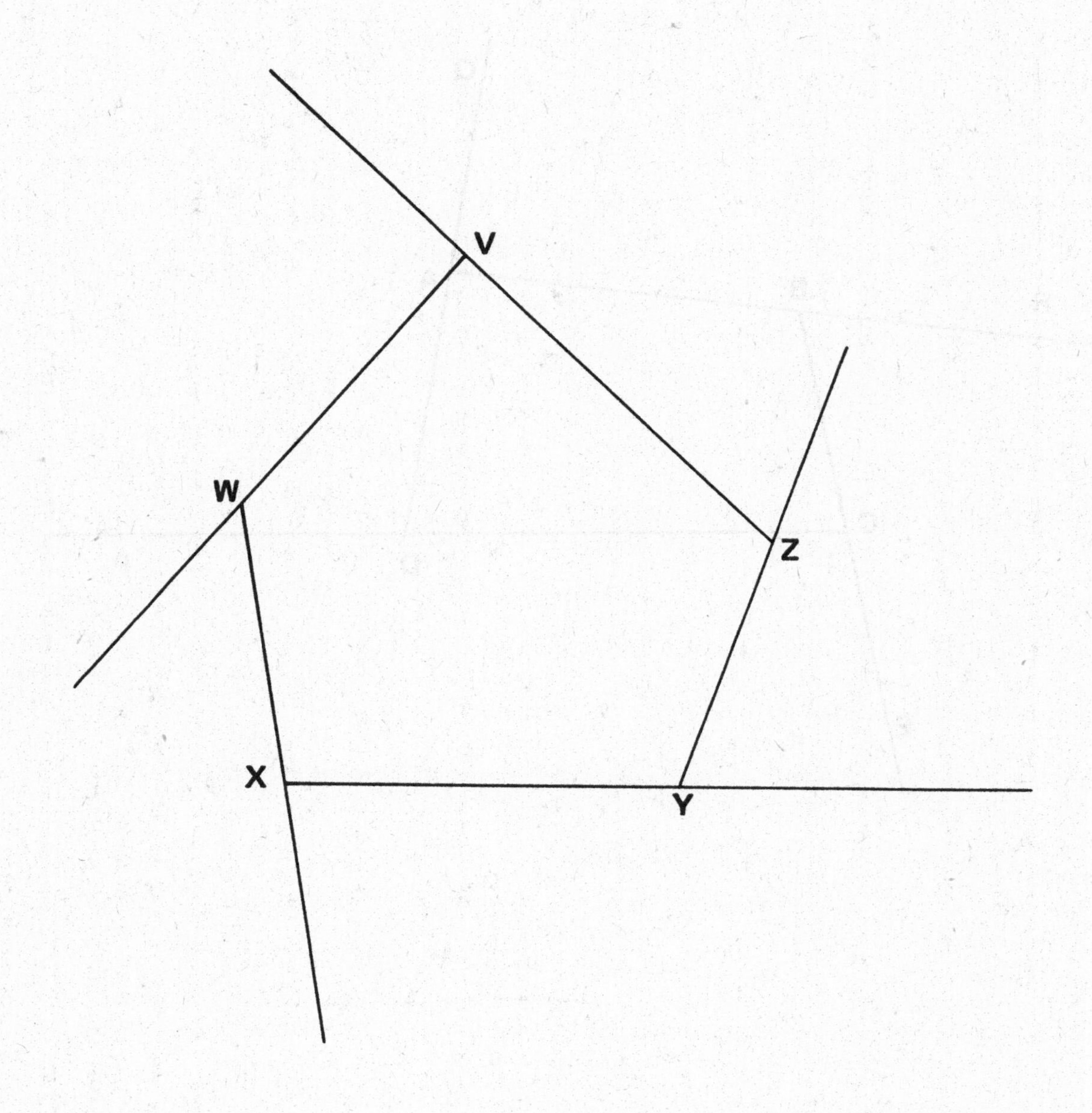

The Sum of the Angles of a Triangle

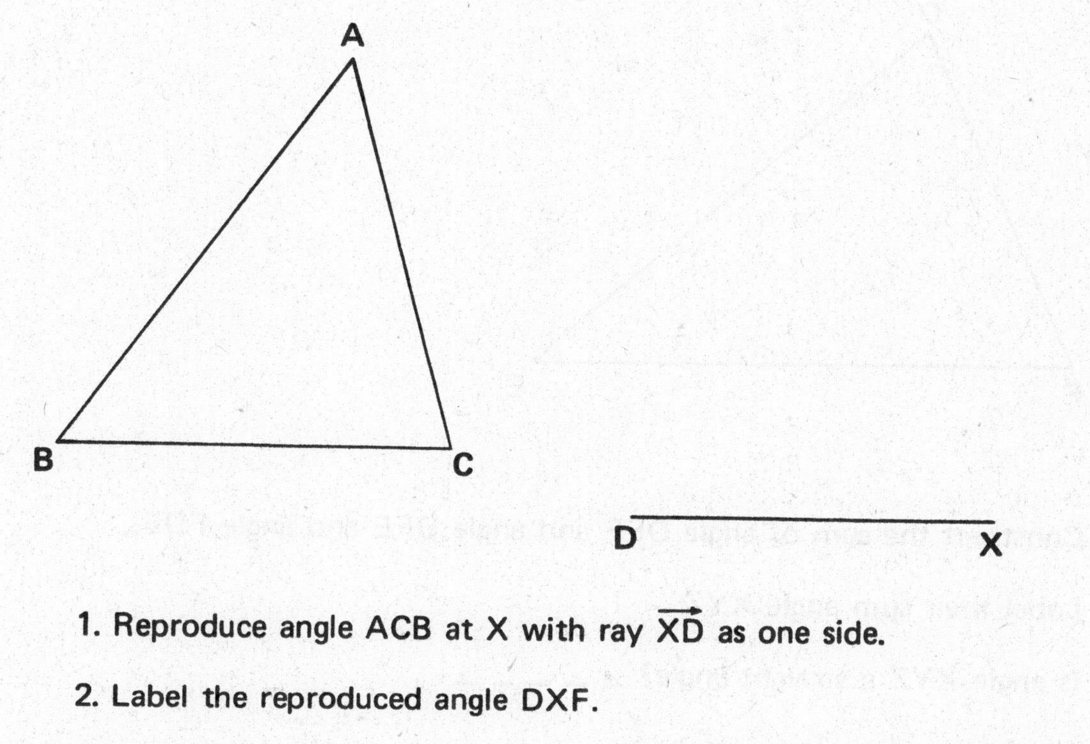

1. Reproduce angle ACB at X with ray $\overrightarrow{XD}$ as one side.
2. Label the reproduced angle DXF.
3. Construct an angle congruent to angle ABC at X with ray $\overrightarrow{XF}$ as one side. Draw its other side to the right of $\overrightarrow{XF}$.
4. Label the reproduced angle FXG.
5. Construct an angle congruent to angle BAC at X with ray $\overrightarrow{XG}$ as one side. Draw its other side to the right of $\overrightarrow{XG}$.
6. Label the angle GXH.
7. Angle DXH is the sum of angle _ _ _ _ _ and angle _ _ _ _ _ and angle _ _ _ _ _.
8. Angle DXH is a _ _ _ _ _ _ _ _ _ _ _ _ _ _ _ _ _ angle.

   (a) right
   (b) straight
   (c) vertical
   (d) exterior

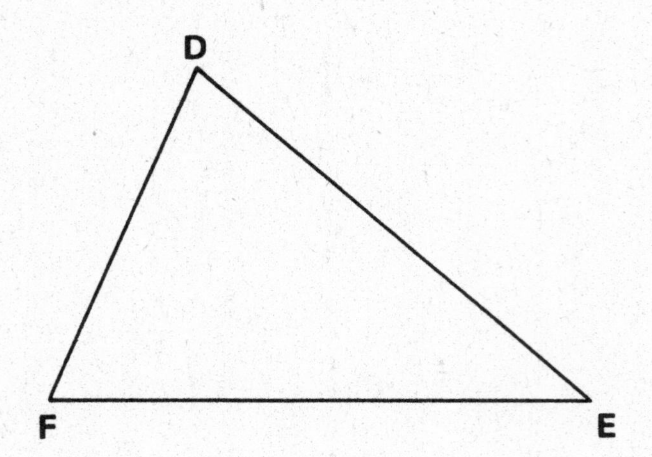

1. Construct the sum of angle DEF and angle DFE and angle FDE.

2. Label their sum angle XYZ.

3. Is angle XYZ a straight angle? _ _ _ _ _

   Why? _

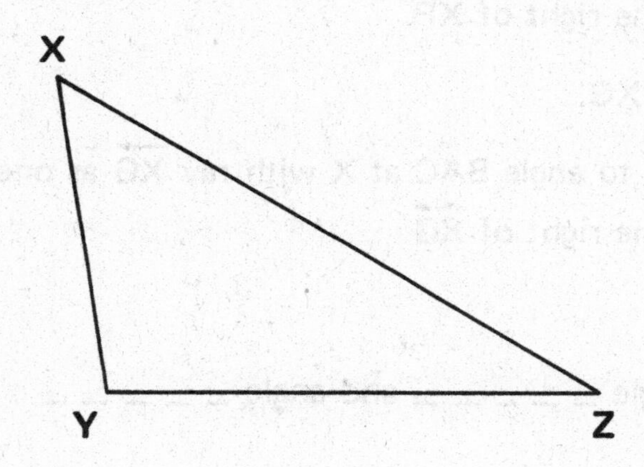

4. Construct the sum of the angles of triangle XYZ.

5. Does the sum form a straight angle? _ _ _ _ _

6. The sum of the angles of a triangle is _ _ _ _ _ _ _ _ _ _ _ _ _ _ a straight angle.

   (a) always (b) sometimes (c) never

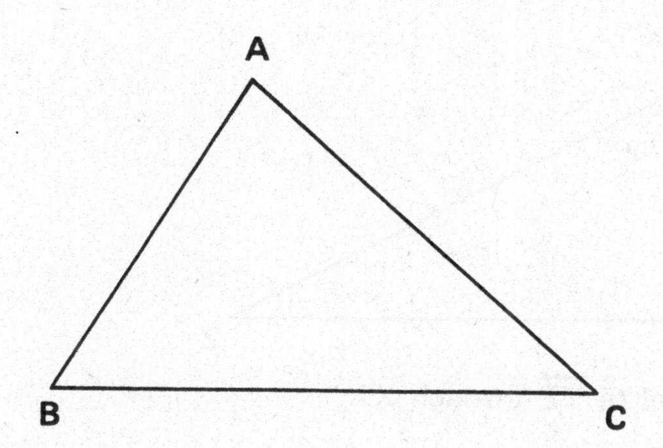

1. Extend side $\overline{BC}$ to the right of C.

   Label as D a point on $\overrightarrow{BC}$ to the right of C.

2. Construct the parallel to $\overleftrightarrow{AB}$ through point C.

3. Label as F a point above C on the parallel.

4. Compare angle FCD and angle ABC.

   Are they congruent? _ _ _ _ _

   Angle FCD and angle ABC are _ _ _ _ _ _ _ _ _ _ _ _ _ _ _ angles.

   (a) vertical
   (b) alternate interior
   (c) corresponding

5. Compare angle ACF and angle BAC.

   Are they congruent? _ _ _ _ _

   Angle ACF and angle BAC are _ _ _ _ _ _ _ _ _ _ _ _ _ _ _ angles.

   (a) vertical
   (b) alternate interior
   (c) corresponding

6. Angle BCD is the sum of angle _ _ _ _ _ and angle _ _ _ _ _ and angle _ _ _ _ _.

7. Is angle BCD a straight angle? _ _ _ _ _

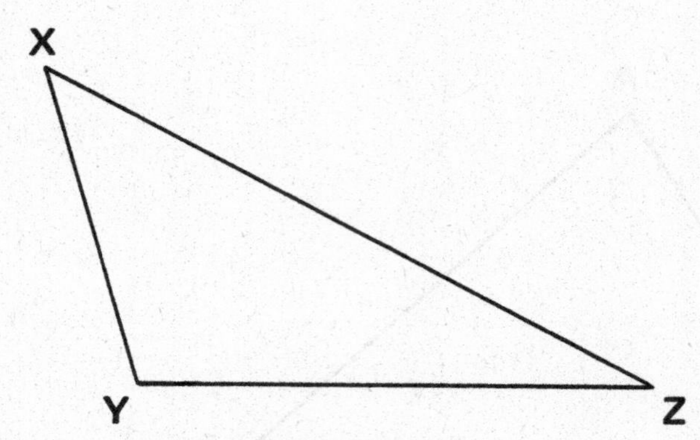

1. Construct the line parallel to $\overleftrightarrow{YZ}$ through X.
2. Label as V a point on the parallel to the left of X.
   Label as W a point on $\overleftrightarrow{VX}$ to the right of X.
3. Extend $\overline{YX}$ beyond X.
   Label as M a point on $\overrightarrow{YX}$ above X.
4. Extend $\overline{ZX}$ beyond X.
   Label as N a point on $\overrightarrow{ZX}$ above X.
5. Angle XYZ is congruent to angle _ _ _ _ _.
   Mark these angles in your figure.

   They are _ _ _ _ _ _ _ _ _ _ _ _ _ _ _ _ _ _ angles.

   (a) vertical (b) corresponding (c) alternate interior
6. Angle YXZ is congruent to angle _ _ _ _ _.
   Mark these angles in your figure.

   They are _ _ _ _ _ _ _ _ _ _ _ _ _ _ _ _ _ _ angles.

   (a) vertical (b) corresponding (c) alternate interior
7. Angle XZY is congruent to angle _ _ _ _ _.
   Mark these angles in your figure.

   They are _ _ _ _ _ _ _ _ _ _ _ _ _ _ _ _ _ _ angles.

   (a) vertical (b) corresponding (c) alternate interior
8. The sum of angle _ _ _ _ _ and angle _ _ _ _ _ and angle _ _ _ _ _ of triangle XYZ is straight angle VXW.

Construct the sum of angle ADC, angle DCB, angle ABC, and angle DAB.

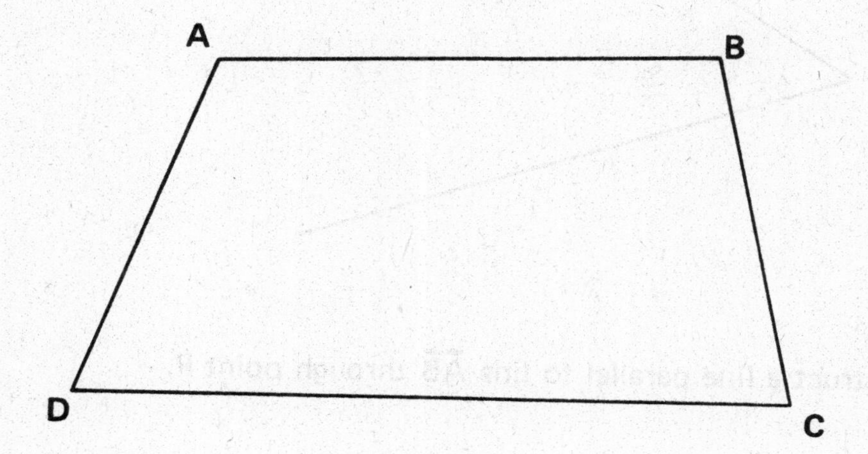

Is the sum a straight angle? _ _ _ _ _

Practice Test

1. Bisect the angle.

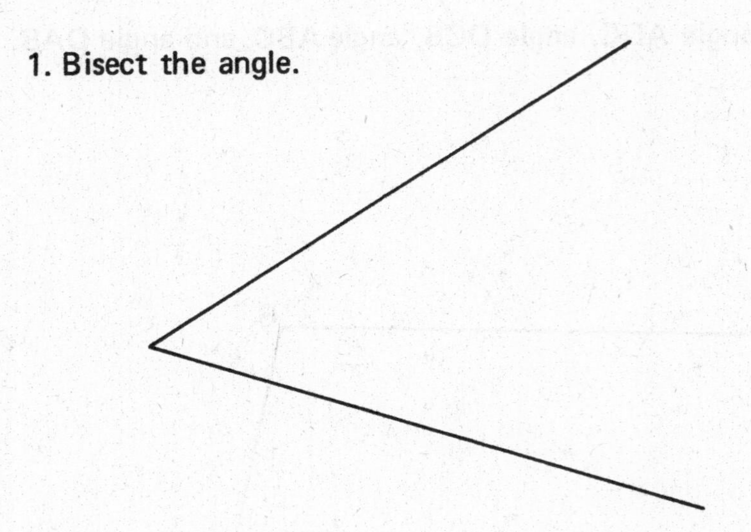

2. Construct a line parallel to line $\overleftrightarrow{AB}$ through point P.

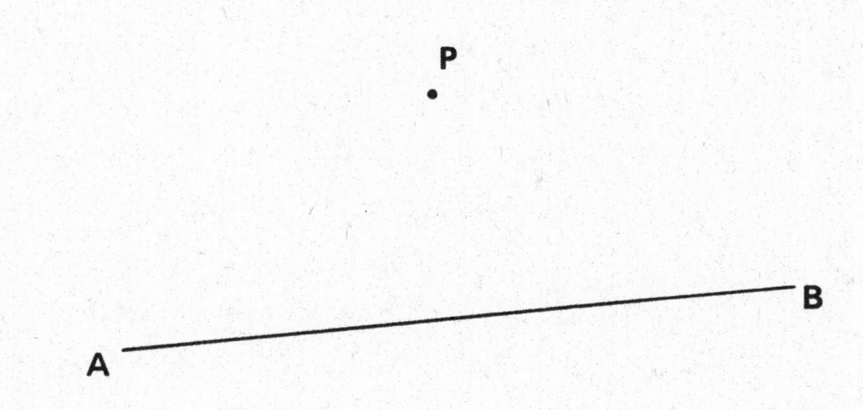

3. Construct an isosceles triangle.

4. Construct a parallelogram with $\overline{AB}$ and $\overline{BC}$ as two sides.

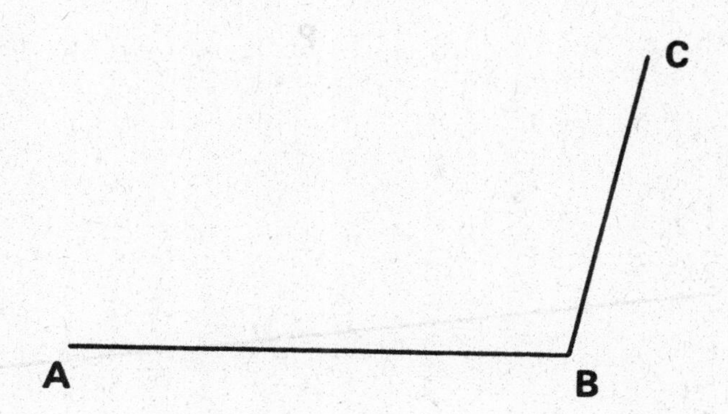

5. Construct the sum of the given angles.

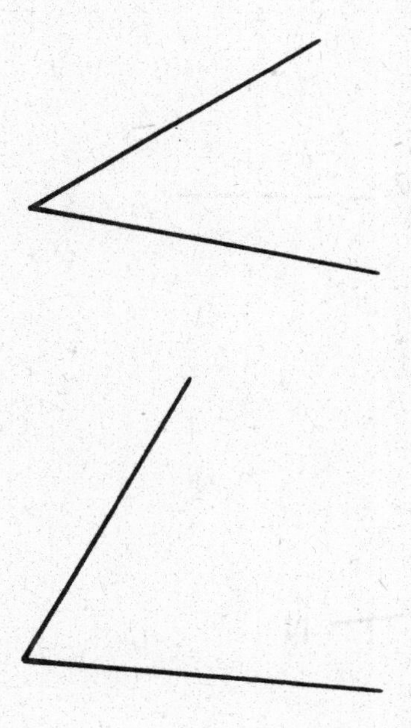

6. Construct a perpendicular to the line through point P.

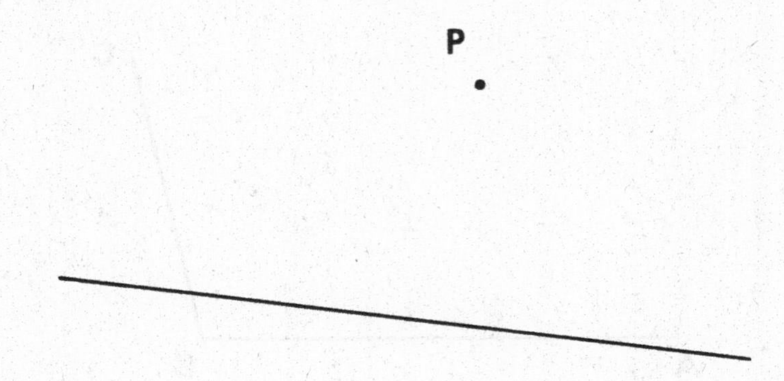

7. Construct the perpendicular bisector of $\overline{AB}$.

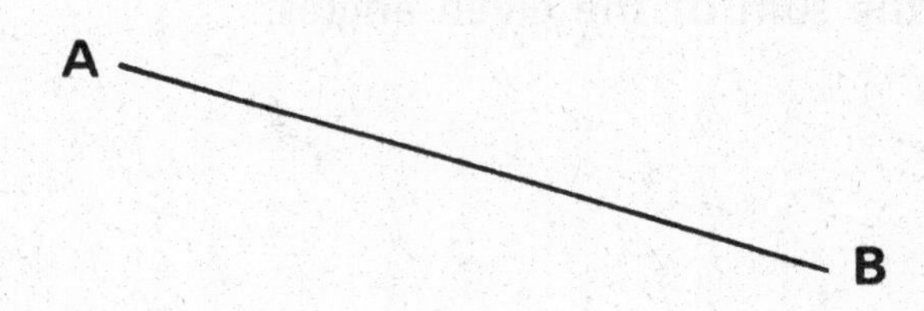

8. Check: Does line $\overleftrightarrow{CD}$ bisect segment $\overline{MN}$? _ _ _ _ _

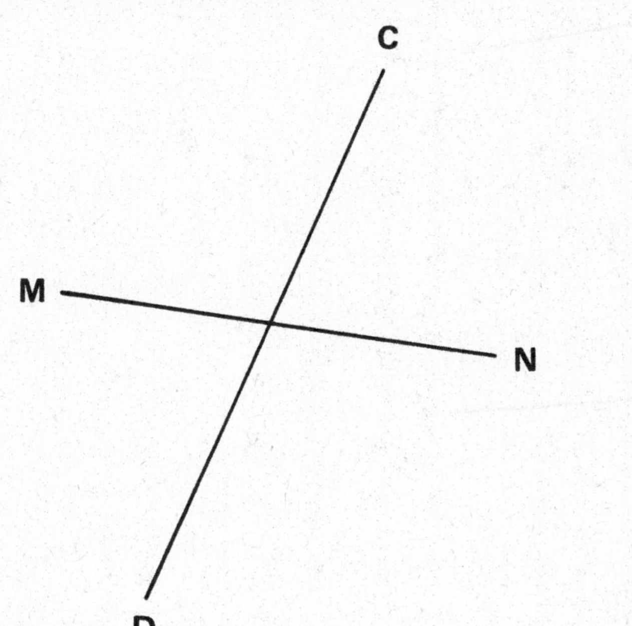

The Chord of a Circle

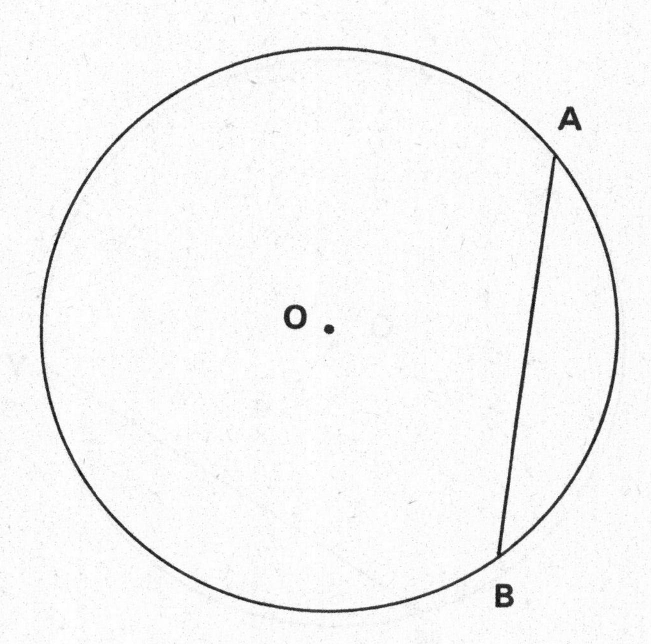

1. A chord of a circle is a segment whose endpoints are on the circle.

   Is $\overline{AB}$ a chord? _ _ _ _ _

2. Draw a chord of the circle and label it $\overline{CD}$.

3. Draw a radius of the circle and label it $\overline{OX}$.

   Draw another radius of the circle and label it $\overline{OZ}$.

4. Compare $\overline{OX}$ and $\overline{OZ}$.

   $\overline{OX}$ is _ _ _ _ _ _ _ _ _ _ _ _ _ _ _ _ _ _ _ $\overline{OZ}$.

   (a) shorter than

   (b) congruent to

   (c) longer than

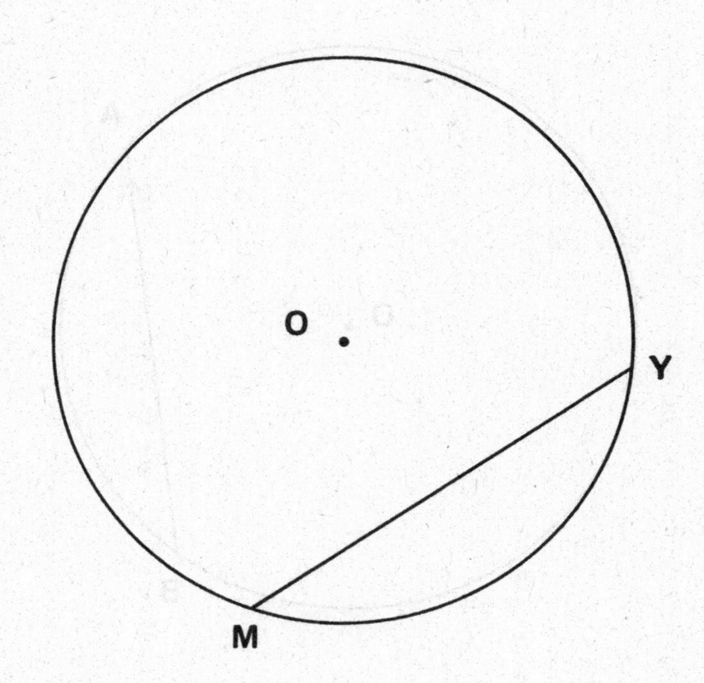

1. O is the center of the circle.

   Construct the perpendicular from O to chord $\overline{MY}$.

2. Label as N the point where the perpendicular intersects $\overline{MY}$.

3. Compare $\overline{MN}$ and $\overline{NY}$.

   $\overline{MN}$ is _ _ _ _ _ _ _ _ _ _ _ _ _ _ _ _ _ _ _ $\overline{NY}$.

   (a) shorter than

   (b) congruent to

   (c) longer than

4. $\overline{ON}$ _ _ _ _ _ _ _ _ _ _ _ _ _ _ _ _ _ _ _ $\overline{MY}$.

   (a) is congruent to

   (b) bisects

   (c) is parallel to

1. Draw a chord of the circle.

   Label its endpoints A and B.

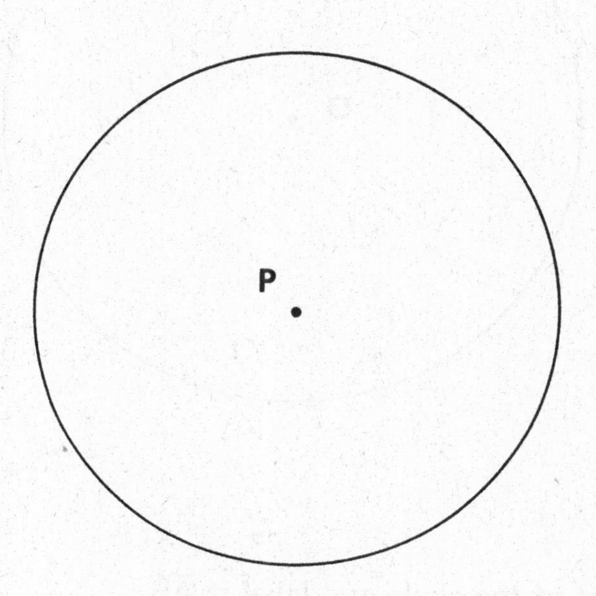

2. Construct the perpendicular from center P to chord $\overline{AB}$.

3. Check: Does the perpendicular bisect segment $\overline{AB}$? _ _ _ _ _

4. If a line from the center of a circle is perpendicular to a chord of the circle, the perpendicular _ _ _ _ _ _ _ _ _ _ _ _ _ _ _ _ bisects the chord.

   (a) always (b) sometimes (c) never

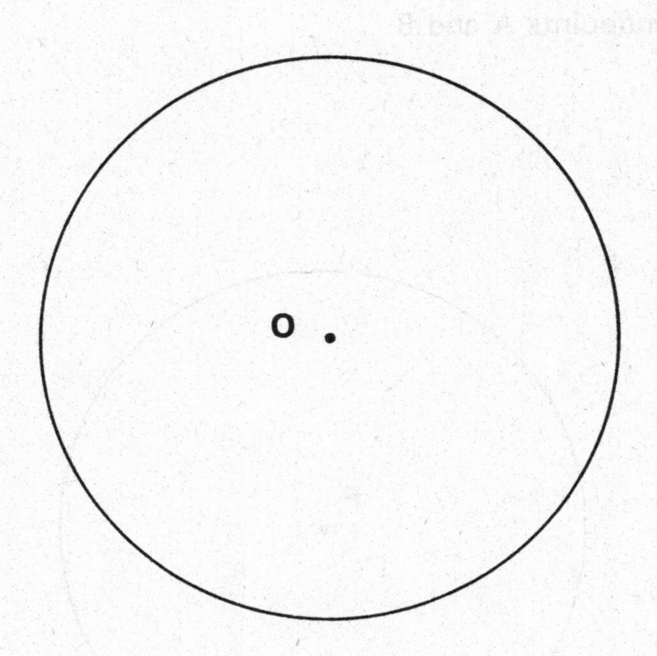

1. Draw a chord of the circle and label it $\overline{AB}$.

2. Construct the perpendicular bisector of $\overline{AB}$.

3. Will the perpendicular bisector of $\overline{AB}$ pass through center O of the circle? _ _ _ _ _

4. The perpendicular bisector of a chord _ _ _ _ _ _ _ _ _ _ _ _ _ _ passes through the center of the circle.

   (a) always (b) sometimes (c) never

Review

1. Construct a perpendicular to the line through the given point.

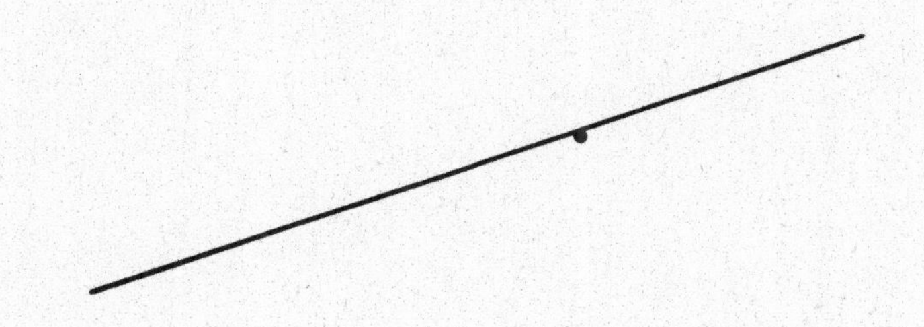

2. Check: Are the lines perpendicular? _ _ _ _ _

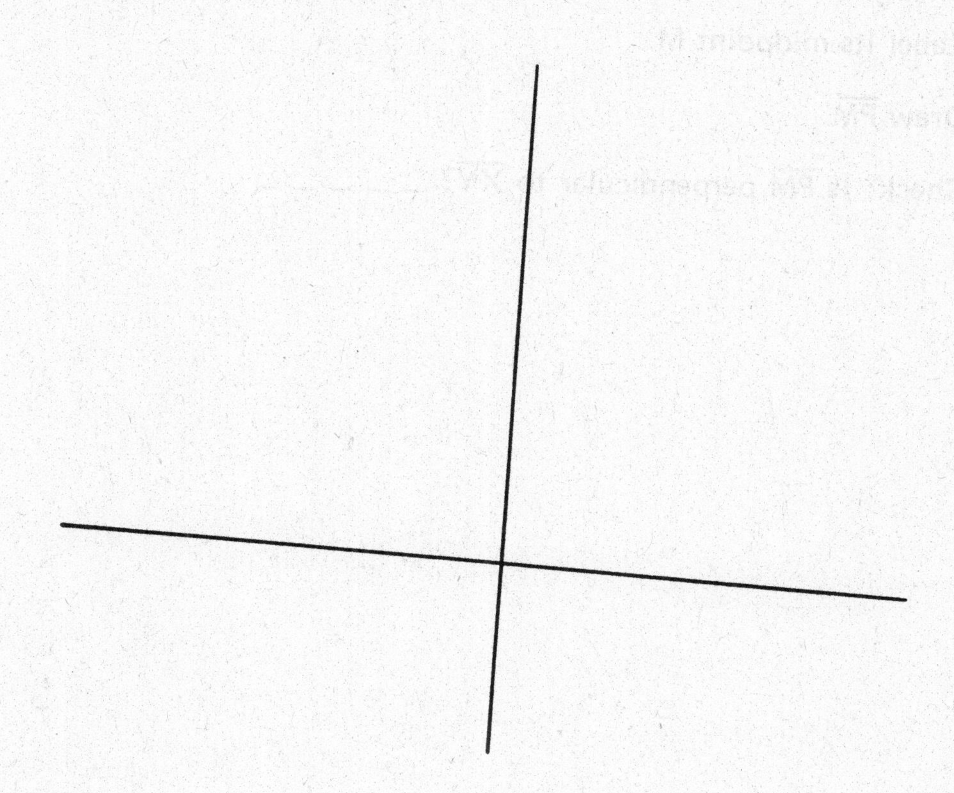

1. Draw a circle and label its center P.

2. Draw a chord of the circle and label it $\overline{XY}$.

3. Bisect $\overline{XY}$.

   Label its midpoint M.

4. Draw $\overline{PM}$.

5. Check: Is $\overline{PM}$ perpendicular to $\overline{XY}$? _ _ _ _ _

Constructing a Circle through the Endpoints of a Segment

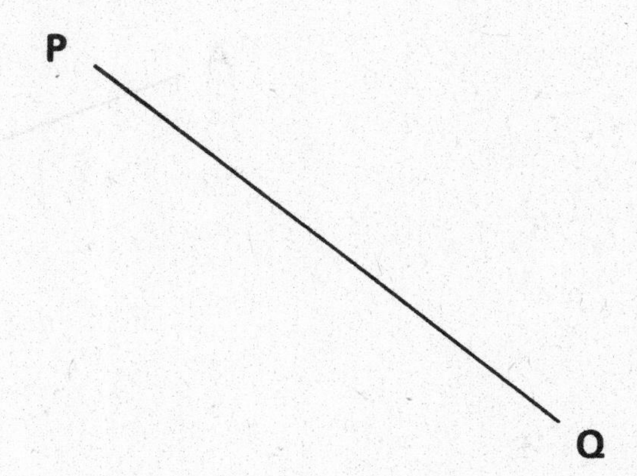

1. Construct the perpendicular bisector of $\overline{PQ}$.

2. Choose a point on the perpendicular bisector and label it O.

3. Draw the circle with center O and radius $\overline{OP}$.

   Does this circle pass through both P and Q? _ _ _ _ _

4. Draw a different circle passing through P and Q.

5. How many circles can you draw through P and Q? _ _ _ _ _

1. Construct a circle passing through the endpoints of segment $\overline{AB}$.

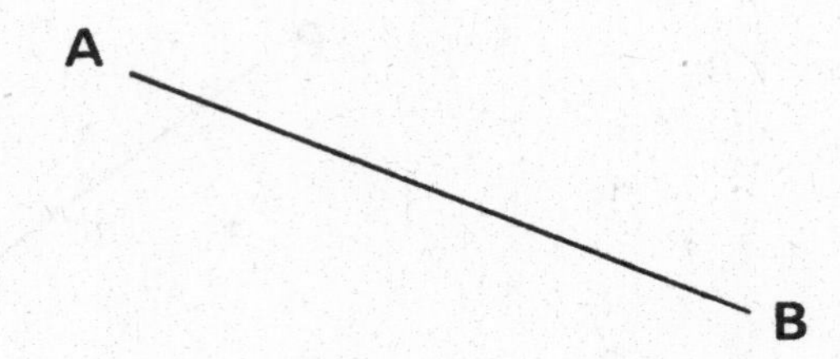

2. Construct a circle passing through the endpoints of both segments.
(Hint: Construct the perpendicular bisectors of both segments.)

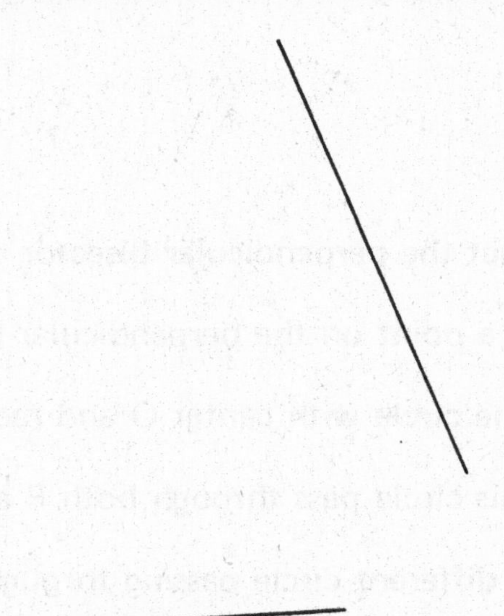

3. A circle can ____________________ be drawn passing through the endpoints of two segments.
(a) always (b) sometimes (c) never

The Tangent to a Circle

A tangent is a line intersecting a circle in exactly one point.

Which of the lines are tangents? ____________________

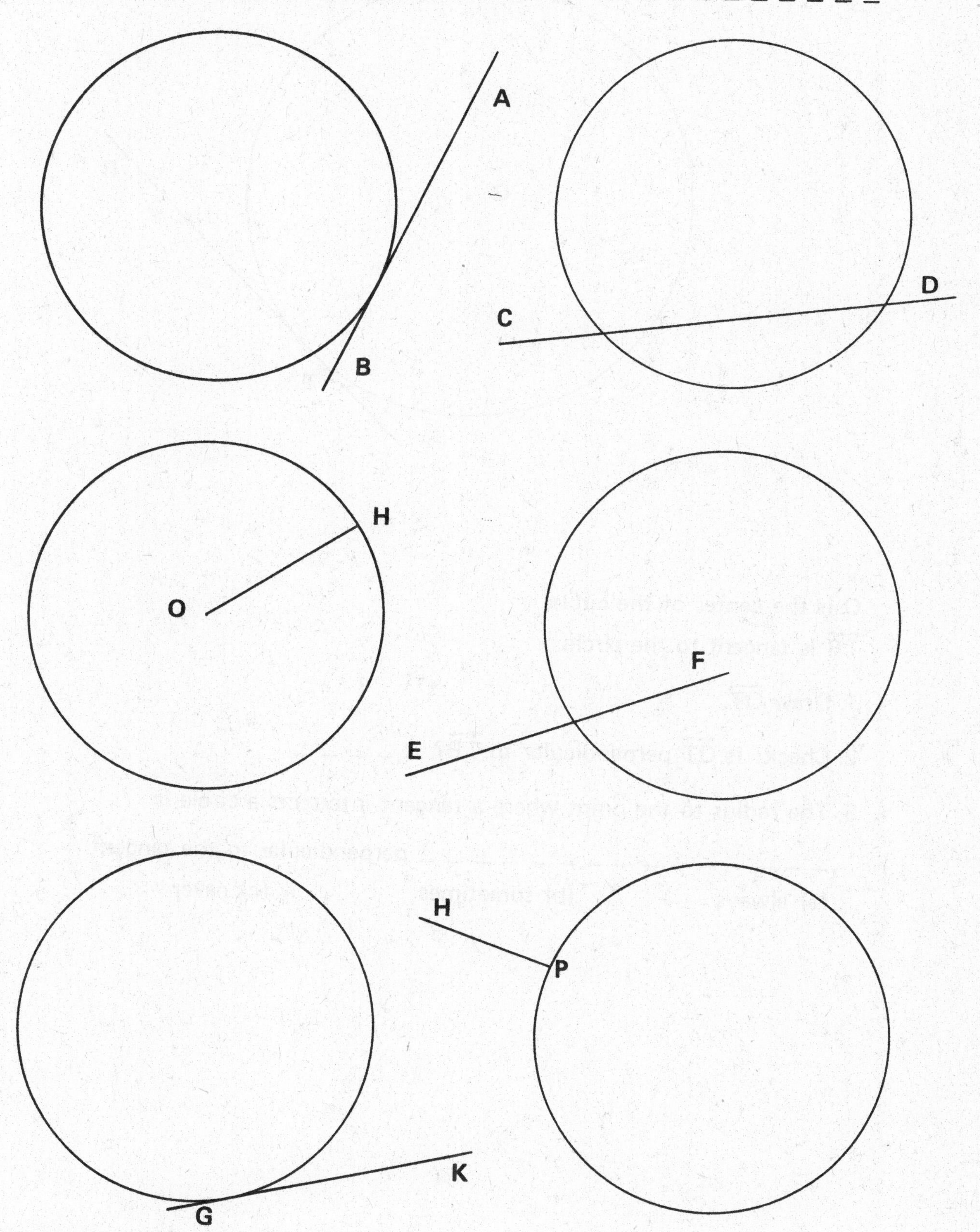

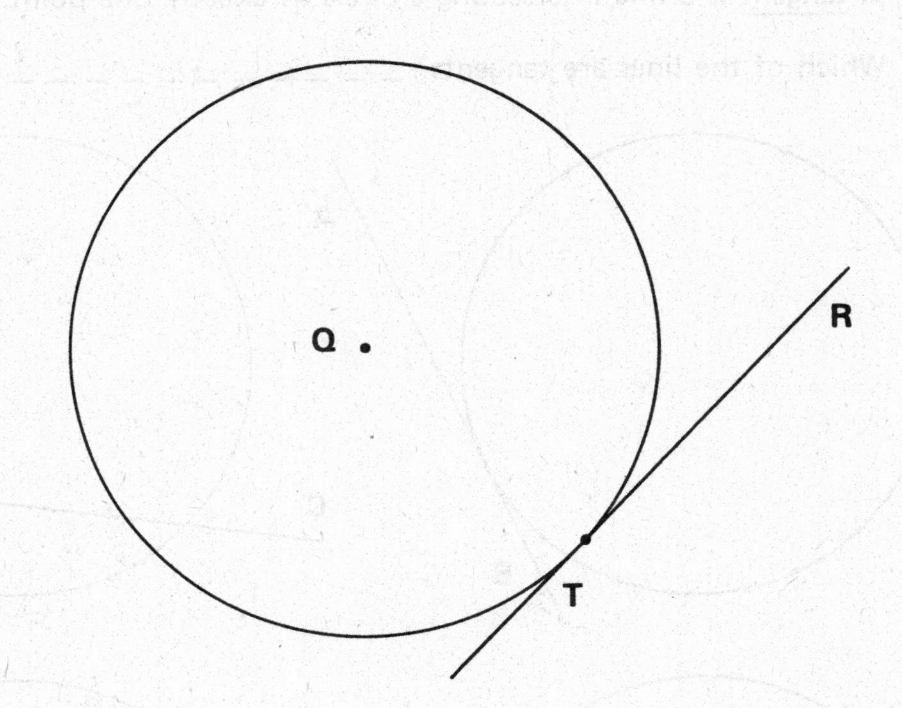

Q is the center of the circle.

$\overleftrightarrow{TR}$ is tangent to the circle.

1. Draw $\overline{QT}$.

2. Check: Is $\overleftrightarrow{QT}$ perpendicular to $\overleftrightarrow{TR}$? _ _ _ _ _

3. The radius to the point where a tangent intersects a circle is

   _ _ _ _ _ _ _ _ _ _ _ _ _ _ _ _ _ perpendicular to the tangent.

   (a) always (b) sometimes (c) never

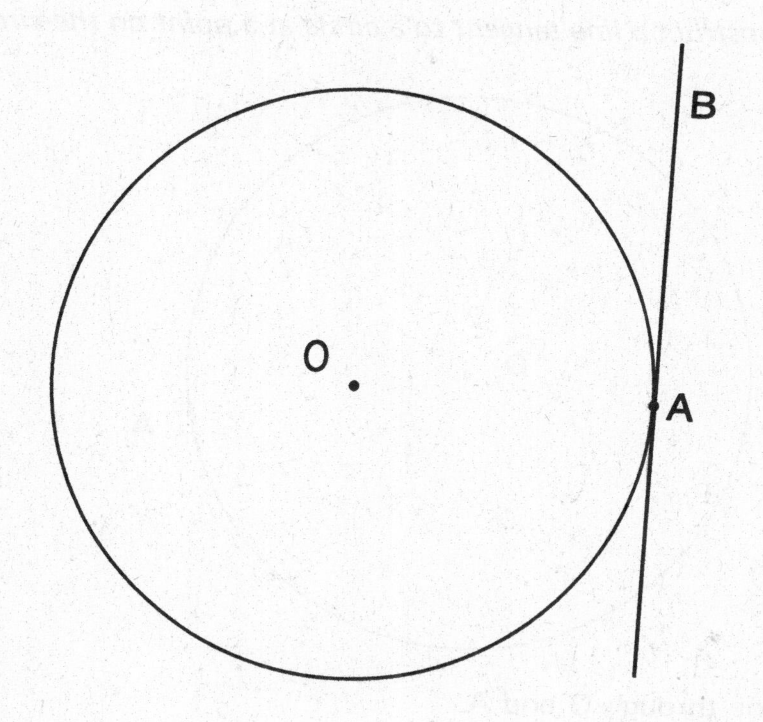

1. Construct the perpendicular to tangent $\overleftrightarrow{AB}$ through point A.
2. Does the perpendicular pass through center O? _ _ _ _ _
3. The perpendicular to a tangent through the point where the tangent intersects the circle will _ _ _ _ _ _ _ _ _ _ _ _ _ _ _ _ pass through the center of the circle.

   (a) always (b) sometimes (c) never

Problem: *Construct a line tangent to a circle at a point on the circle.*

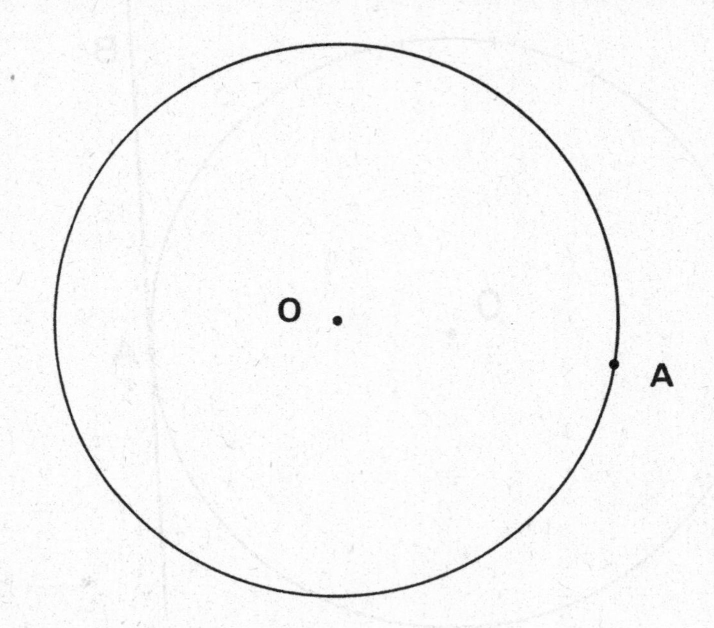

Solution:

1. Draw a line through O and A.

2. Construct the perpendicular to line $\overleftrightarrow{OA}$ through point A.

3. The perpendicular intersects the circle in _ _ _ _ _ _ _ _ _ _ _ _ _ .

   (a) one point (c) three points

   (b) two points (d) four points

4. Is the perpendicular a tangent to the circle? _ _ _ _ _

5. Construct a line perpendicular to radius $\overline{QP}$ at point P.

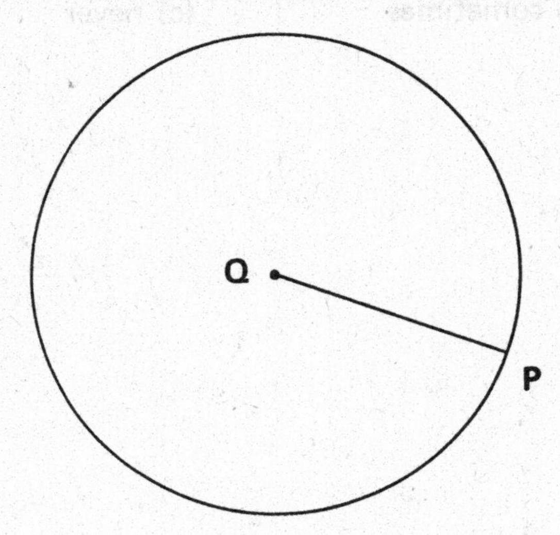

6. Is the line tangent to the circle? _ _ _ _ _ _ _ _ _ _ _ _ _ _ _ _ _

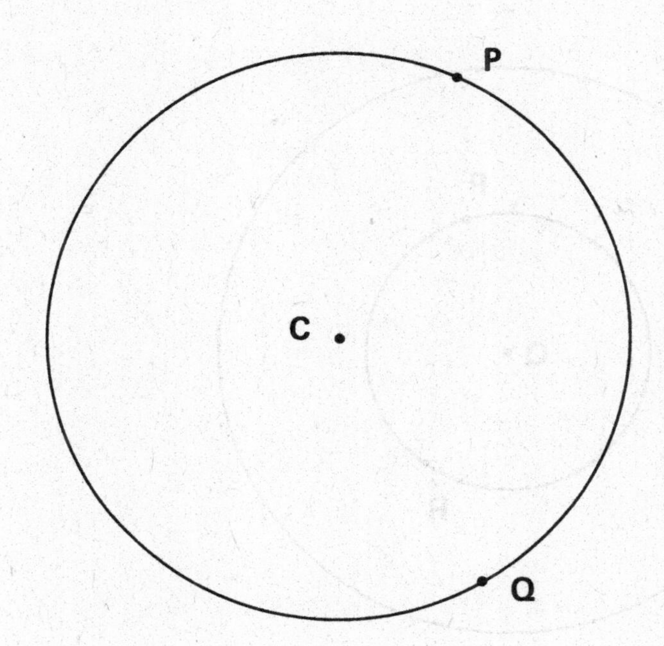

1. Construct a line tangent to the circle at point P.

2. Construct a line tangent to the circle at point Q.

3. Make the tangent lines intersect.
   Label the point of intersection T.

4. Compare $\overline{PT}$ and $\overline{QT}$.

   $\overline{PT}$ is _ _ _ _ _ _ _ _ _ _ _ _ _ _ _ _ _ _ $\overline{QT}$.

   (a) shorter than

   (b) congruent to

   (c) longer than

Q is the center of both circles.

1. Construct the line tangent to the smaller circle at point P.

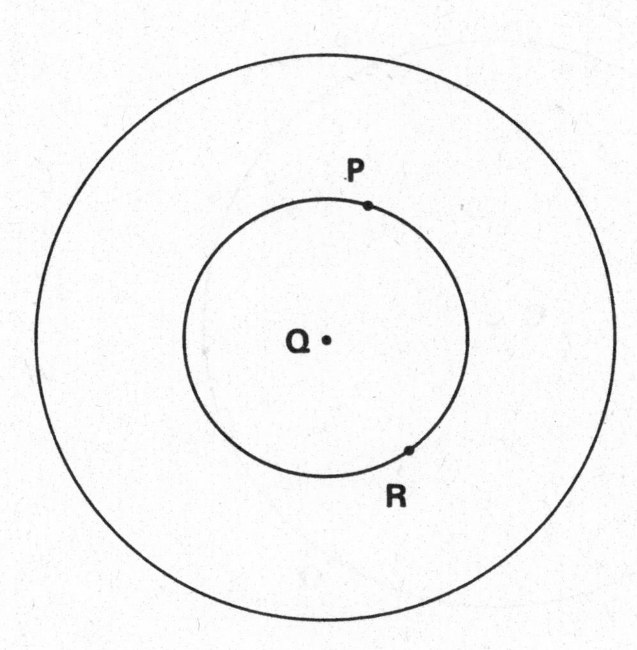

2. Label as A and B the points where it intersects the larger circle.

3. Construct the line tangent to the smaller circle at point R.

4. Label as C and D the points where it intersects the larger circle.

5. Compare $\overline{AB}$ and $\overline{CD}$.

$\overline{AB}$ is _ _ _ _ _ _ _ _ _ _ _ _ _ _ _ _ _ _ $\overline{CD}$.

(a) shorter than

(b) congruent to

(c) longer than

Constructing a Circle Tangent to a Line

Problem: *Construct a circle tangent to a given line at a point.*

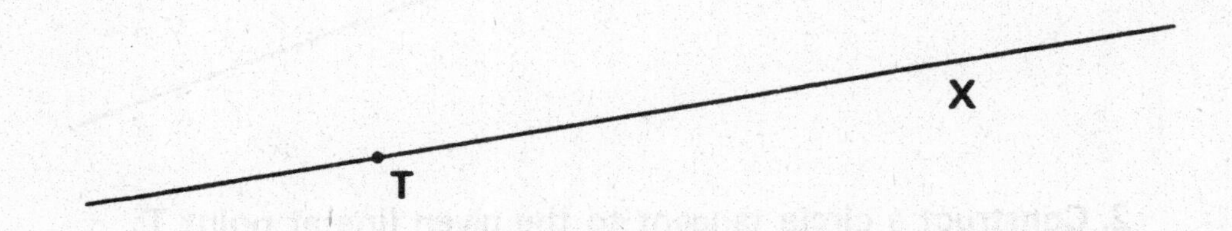

Solution:

1. Construct a perpendicular to line $\overleftrightarrow{TX}$ through point T.

2. Choose a point on the perpendicular above T and label it Q.

3. Draw the circle with center Q and radius $\overline{QT}$.

4. The circle intersects line $\overleftrightarrow{TX}$ in _ _ _ _ _ _ _ _ _ _ _ _ _ _ _ _ _.

   (a) one point

   (b) two points

   (c) three points

   (d) four points

5. Is the circle tangent to line $\overleftrightarrow{TX}$? _ _ _ _ _

   Why? _

6. Draw another circle tangent to line $\overleftrightarrow{TX}$ at point T.

Review

1. Bisect the given angle.

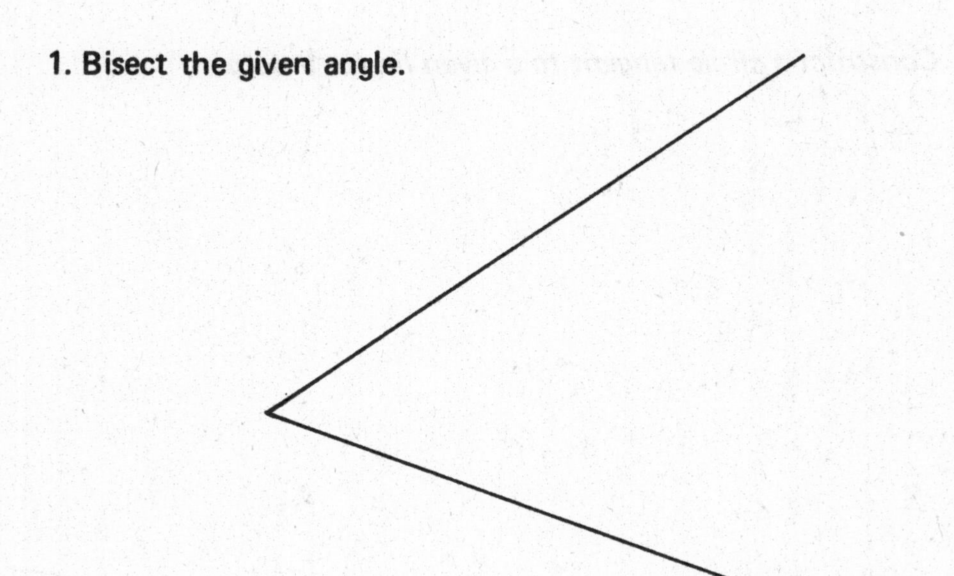

2. Construct a circle tangent to the given line at point T.

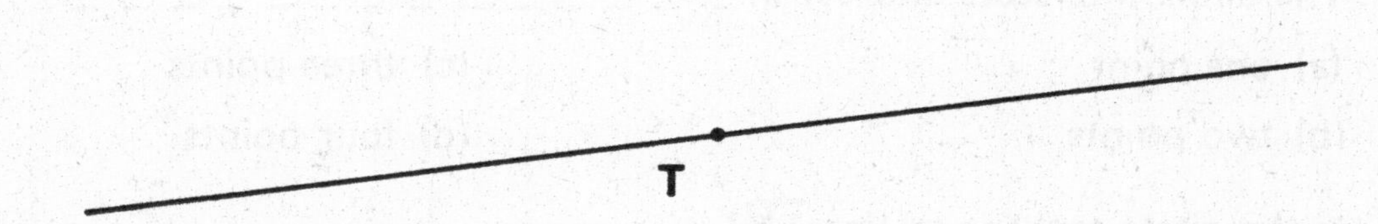

3. Construct a perpendicular to the line from the given point.

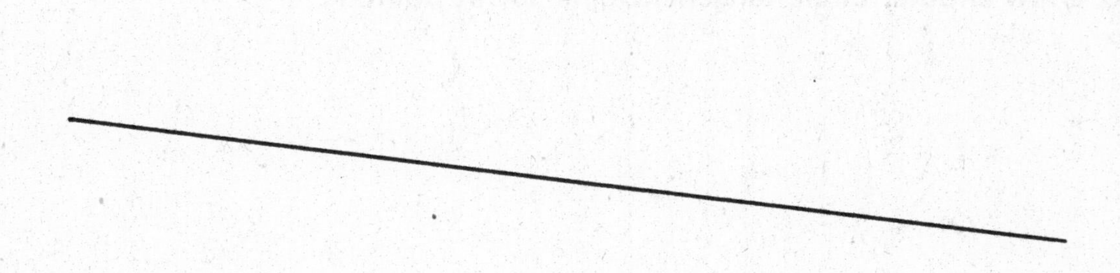

Problem: *Construct a circle tangent to both sides of a given angle.*

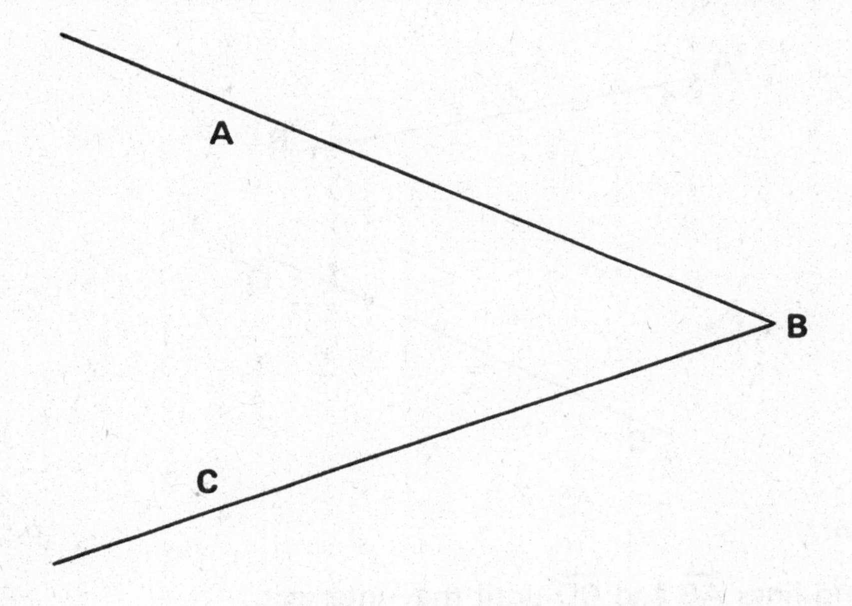

Solution:

1. Bisect angle ABC.

2. Choose a point on the bisector and label it O.

3. Construct the perpendicular from point O to side $\overrightarrow{BC}$.

4. Label as T the point where the perpendicular intersects $\overleftrightarrow{BC}$.

5. Draw the circle with center O and radius $\overline{OT}$.

6. Is the circle tangent to $\overleftrightarrow{BC}$? _ _ _ _ _

   Is the circle tangent to $\overleftrightarrow{AB}$? _ _ _ _ _

Problem: *Construct a circle tangent to two given lines.*

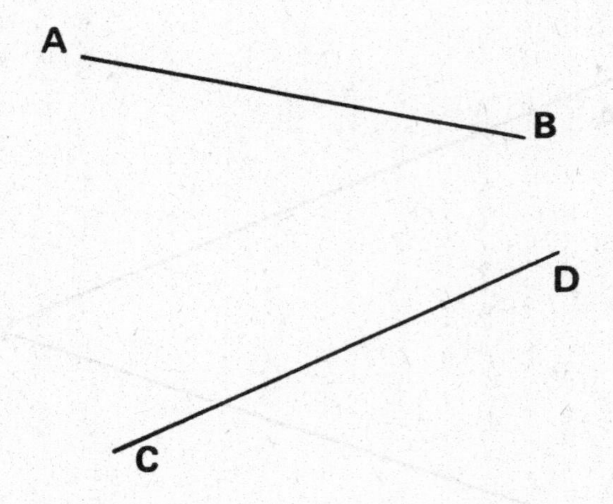

Solution:

1. Extend lines $\overleftrightarrow{AB}$ and $\overleftrightarrow{CD}$ until they intersect.
2. Label the intersection X.
3. Construct a circle tangent to both sides of angle AXC.
4. Is the circle tangent to both lines $\overleftrightarrow{AB}$ and $\overleftrightarrow{CD}$? _ _ _ _ _
5. Construct a circle tangent to both lines.

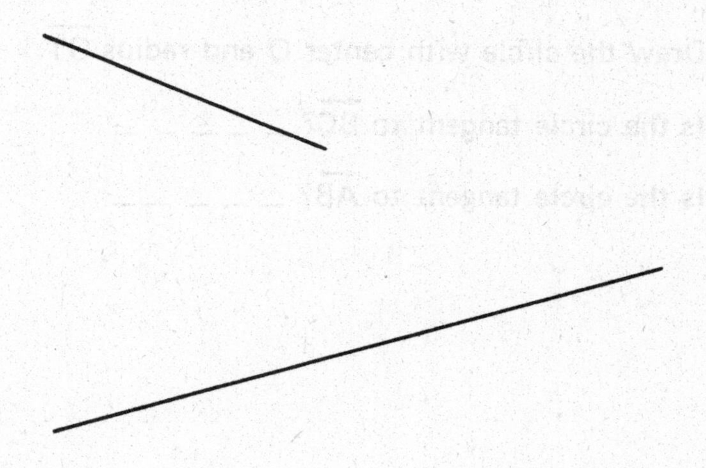

6. How many circles can you construct which are tangent to both lines? _ _ _ _ _

1. Construct a circle tangent to both lines.

2. Construct a circle tangent to both parallel lines.

Tangent Circles

Two circles are tangent if they intersect in exactly one point.

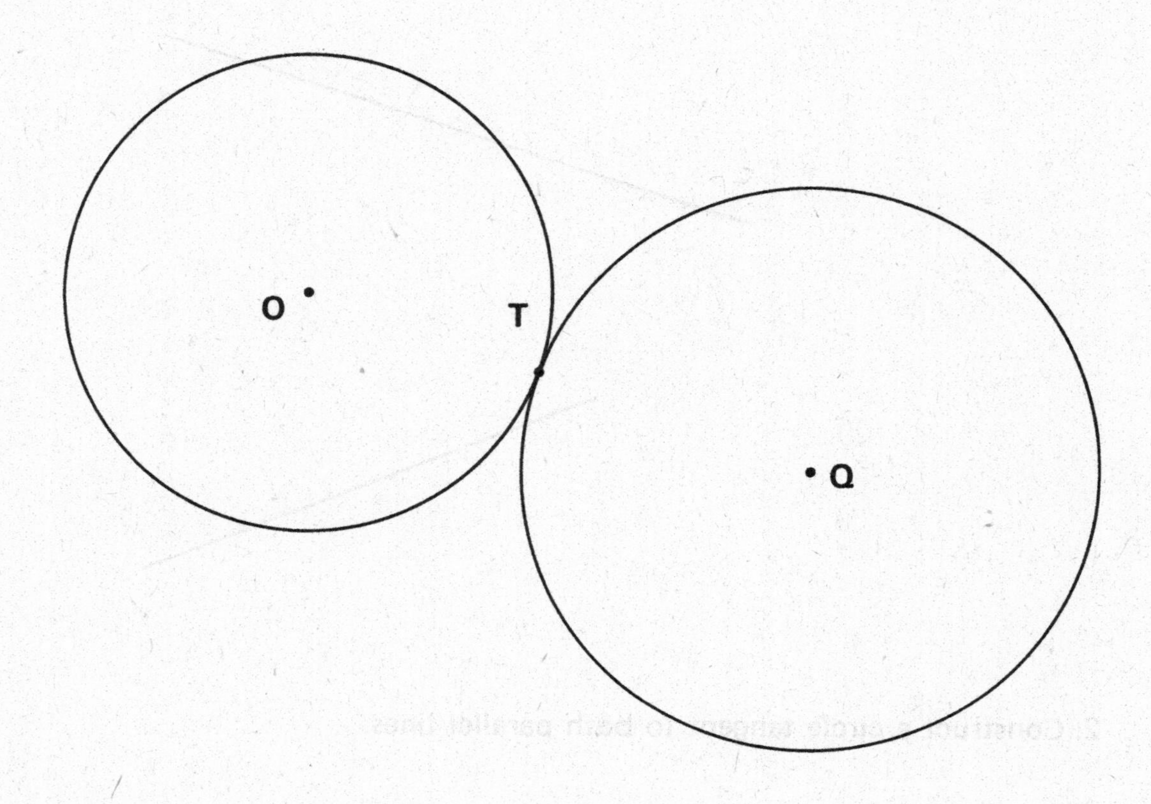

O and Q are the centers of the circles.

1. Draw $\overline{OQ}$.

2. Does $\overline{OQ}$ pass through T? _ _ _ _ _

3. Construct the perpendicular to $\overleftrightarrow{OQ}$ at T.

4. Is the perpendicular tangent to the circle with center O? _ _ _ _ _

5. Is the perpendicular tangent to the circle with center Q? _ _ _ _ _

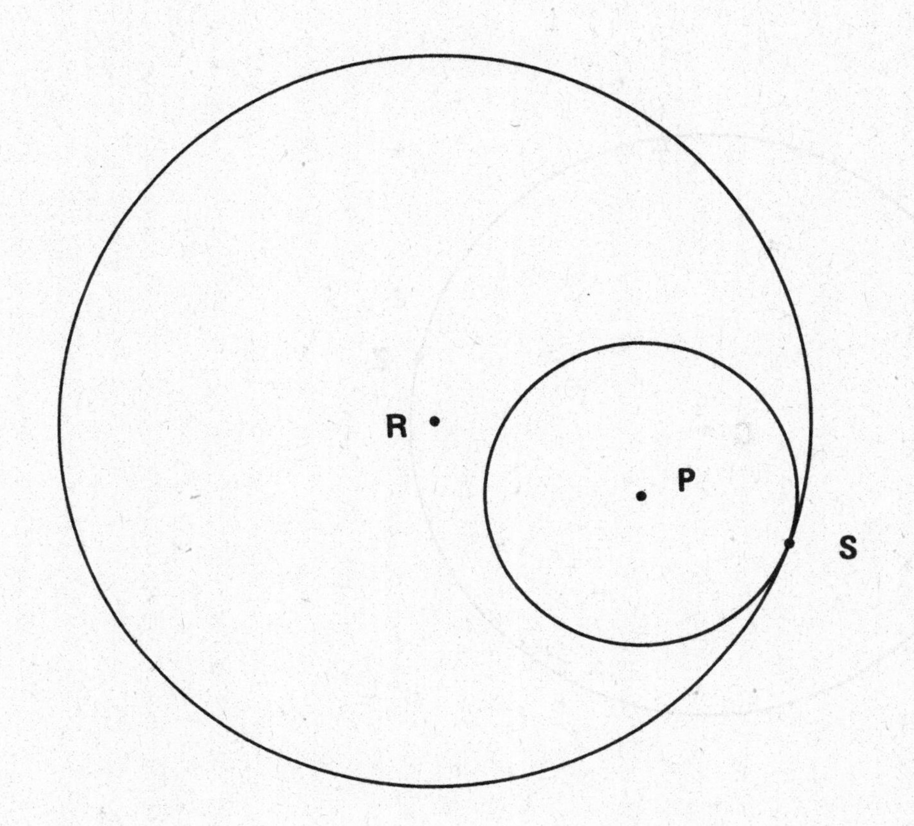

1. Are the circles tangent? _ _ _ _ _

2. Draw the line through R and P.

3. Does it pass through S? _ _ _ _ _

4. Construct the perpendicular to $\overleftrightarrow{RP}$ through point S.

5. Is the perpendicular tangent to both circles? _ _ _ _ _

Problem: *Construct a circle tangent to a given circle at a given point.*

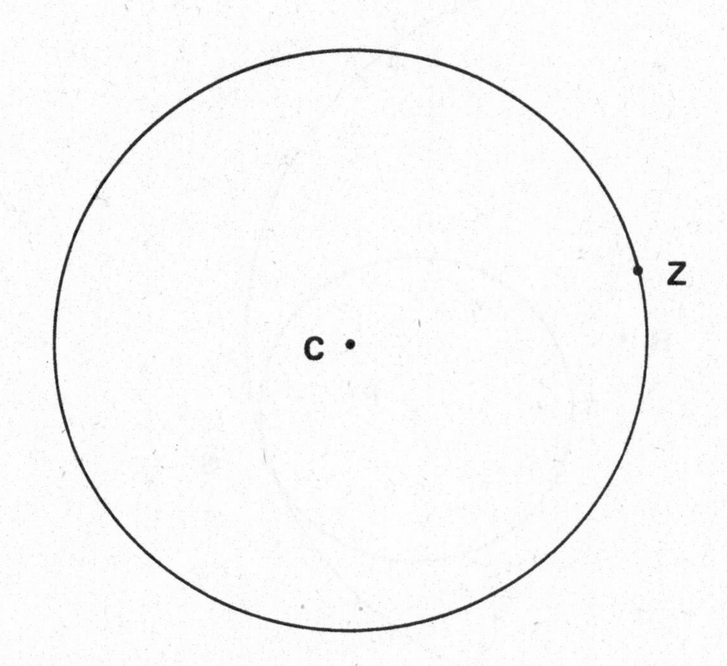

Solution:

C is the center of the circle.

1. Draw a line through C and Z.
2. Choose a point on the line outside the circle.
   Label it O.
3. Draw the circle with center O and radius $\overline{OZ}$.
4. Is this circle tangent to circle C at point Z? _ _ _ _ _
5. Choose a point on line $\overleftrightarrow{CZ}$ inside circle C and label it P.
6. Draw the circle with center P and radius $\overline{PZ}$.
7. How many circles are tangent to each other? _ _ _ _ _

Practice Test

1. Construct a circle tangent to circle Q at P.

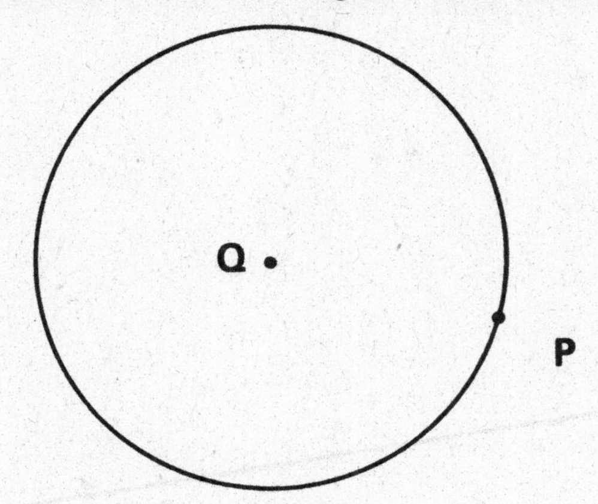

2. Check: Are the angles congruent? _ _ _ _ _

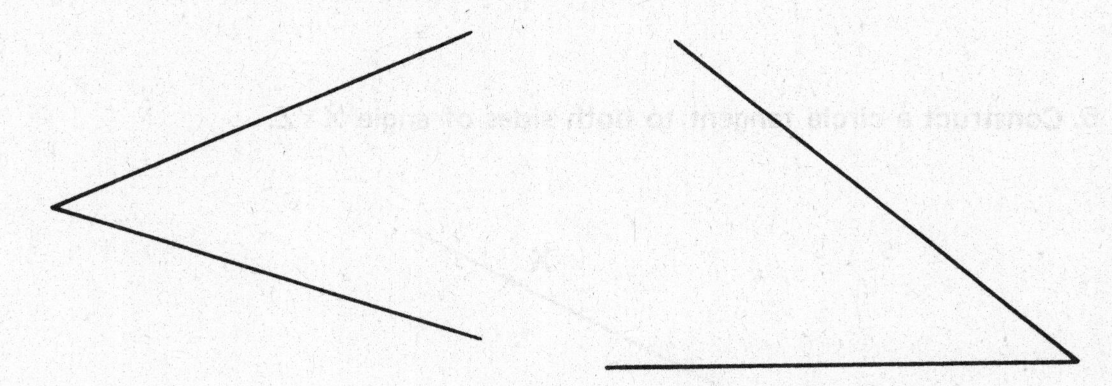

3. Construct a line parallel to the given line through point X.

4. Construct a circle tangent to line $\overleftrightarrow{CD}$ at C.

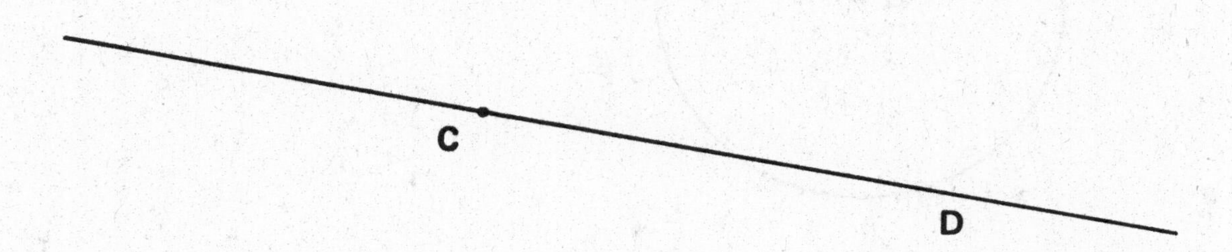

5. Construct a circle tangent to both sides of angle XYZ.

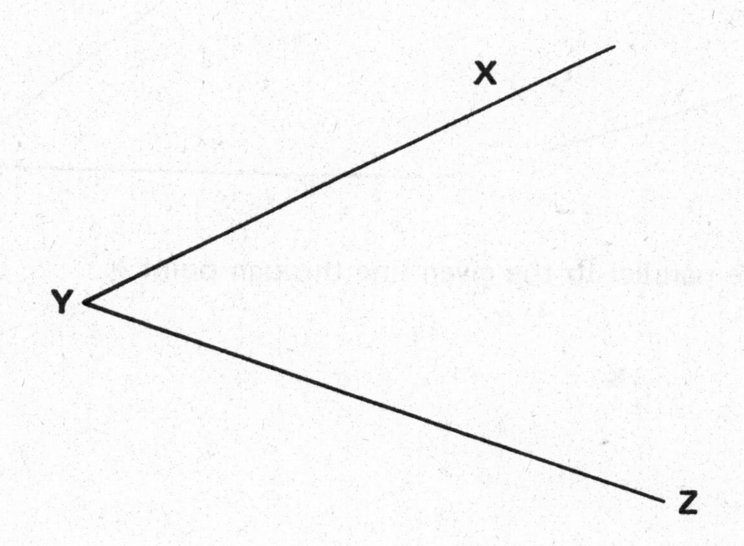

6. Construct a circle passing through the endpoints of segment $\overline{AB}$.

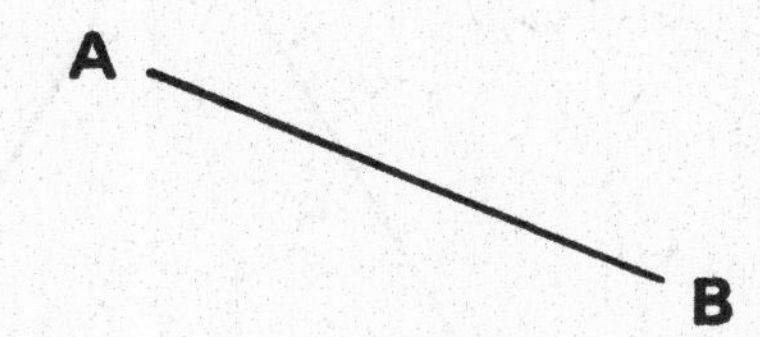

7. Construct a parallelogram with sides $\overline{WY}$ and $\overline{YX}$.

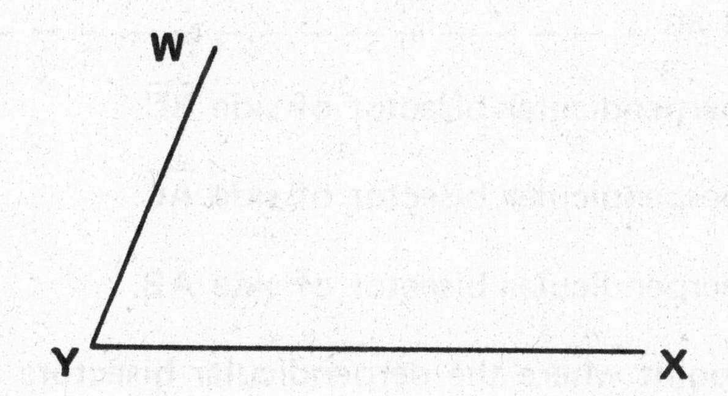

8. Construct the perpendicular to the line through point P.

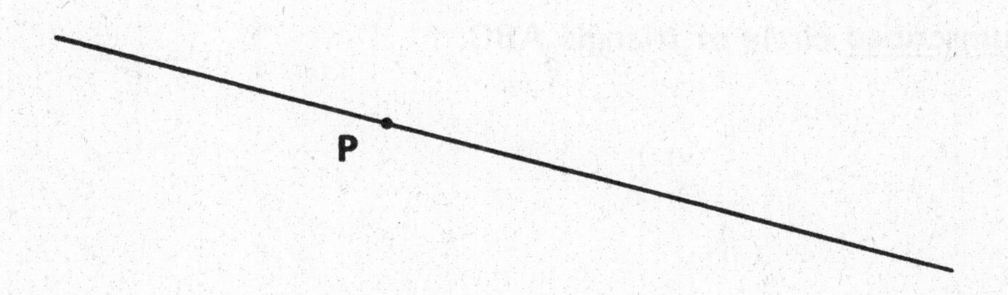

Circumscribing a Circle about a Triangle

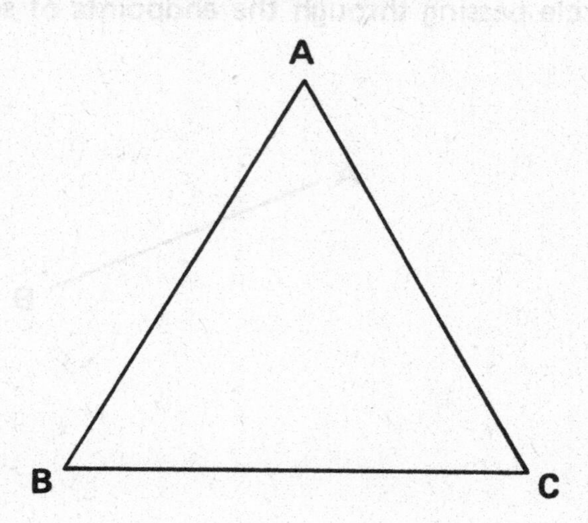

1. Compare sides $\overline{AB}$, $\overline{AC}$, and $\overline{BC}$.

   Are they congruent? _ _ _ _ _

2. Triangle ABC is an _ _ _ _ _ _ _ _ _ _ _ _ _ _ _ _ _ triangle.

3. Construct the perpendicular bisector of side $\overline{BC}$.

4. Construct the perpendicular bisector of side $\overline{AC}$.

5. Construct the perpendicular bisector of side $\overline{AB}$.

6. Label as P the point where the perpendicular bisectors intersect.

7. Draw the circle with center P which passes through A.

8. Does the circle pass through B and C? _ _ _ _ _

9. If the circle passes through all three points A, B, and C, it is the <u>circumscribed</u> circle of triangle ABC.

1. Is triangle ABC an equilateral triangle? _ _ _ _ _

   Why? _

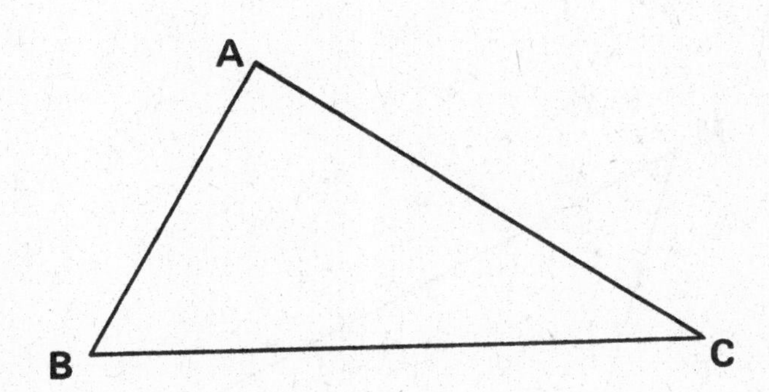

2. Construct the perpendicular bisector of each side of triangle ABC.
3. Label as Q the point where the perpendicular bisectors intersect.
4. Do you think the circle with center Q which passes through A will

   pass through B and C? _ _ _ _ _
5. Draw the circle with center Q passing through A.
6. Is this circle the circumscribed circle of triangle ABC? _ _ _ _ _

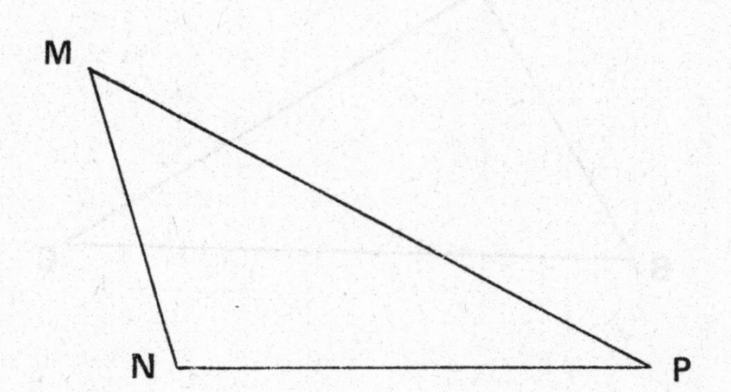

1. Construct the perpendicular bisector of each side of triangle MNP.

2. Label as O the point where the perpendicular bisectors intersect.

3. Compare $\overline{OM}$, $\overline{ON}$, and $\overline{OP}$.

   Are they all congruent? _ _ _ _ _

4. Do you think the circle with center O passing through M will pass through N and P? _ _ _ _ _

   Why? _

5. Draw the circle with center O passing through M.

6. A circle can _ _ _ _ _ _ _ _ _ _ _ _ _ _ be circumscribed around a triangle.

   (a) always (b) sometimes (c) never

1. Circumscribe a circle about the triangle.

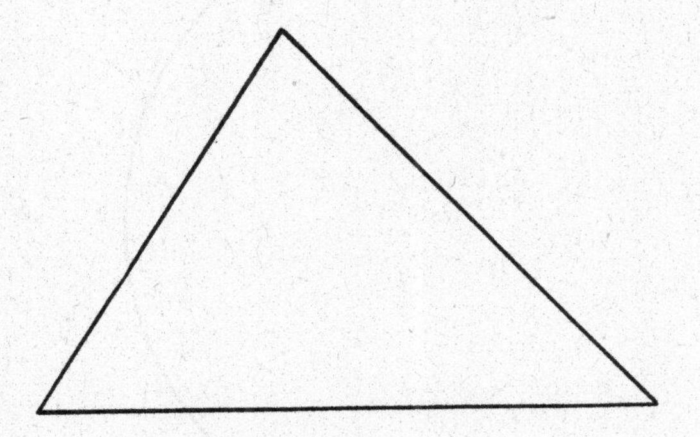

2. Circumscribe a circle about the square.

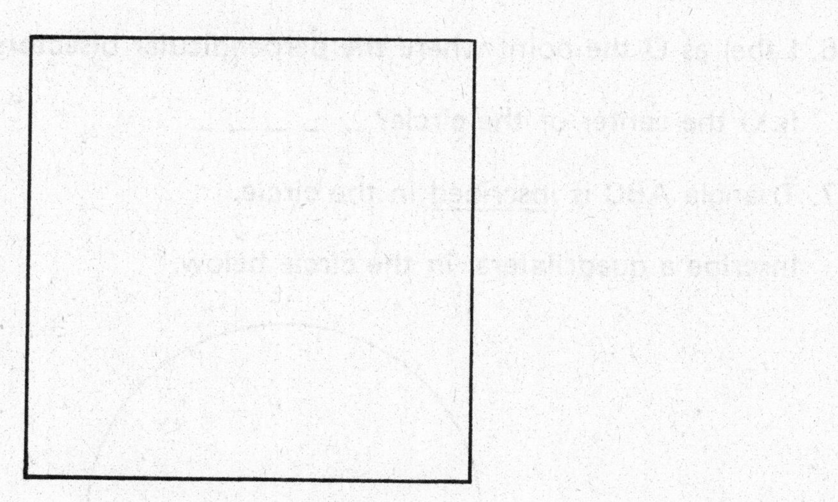

Inscribing a Triangle in a Circle

Problem: *Find the center of a circle.*

Solution:

1. Choose three points on the circle and label them A, B, C.
2. Draw triangle ABC.
3. Construct the perpendicular bisector of $\overline{AB}$.
4. Construct the perpendicular bisector of $\overline{BC}$.
5. Construct the perpendicular bisector of $\overline{AC}$.

   Do all the perpendicular bisectors meet in a point? _ _ _ _ _

6. Label as O the point where the perpendicular bisectors meet.

   Is O the center of the circle? _ _ _ _ _

7. Triangle ABC is inscribed in the circle.

   Inscribe a quadrilateral in the circle below.

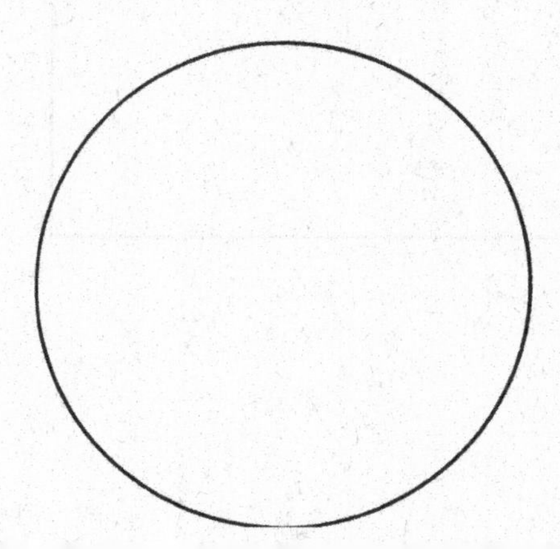

1. Find the center of the circle.

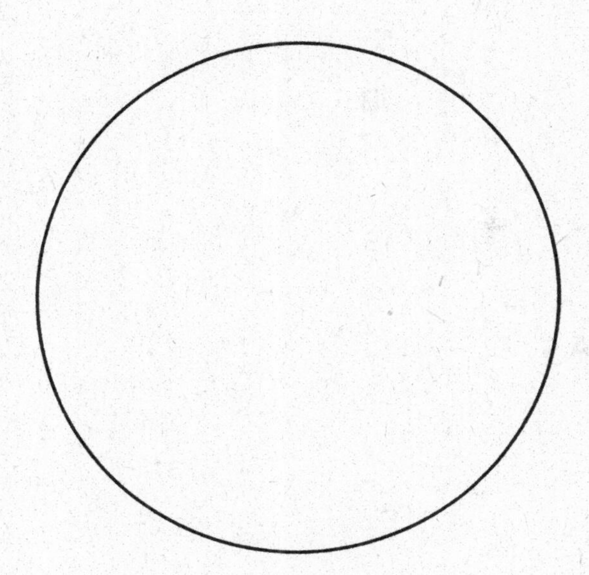

2. Inscribe a quadrilateral in the circle below.

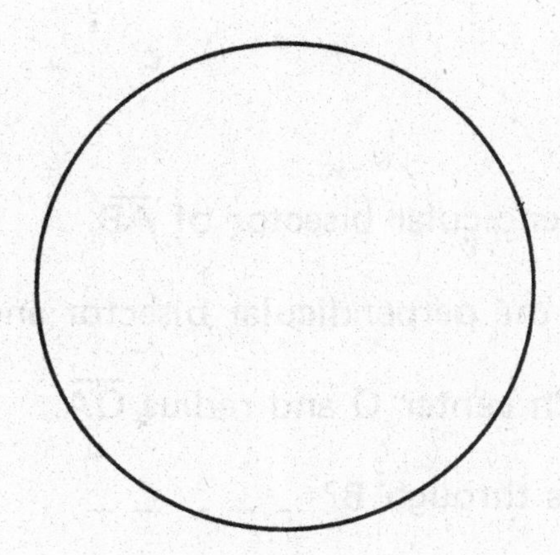

3. Construct the perpendicular bisector of each side.

4. Do the perpendicular bisectors meet in one point? _ _ _ _ _

Label the center of the circle X.

Drawing a Circle through Three Points

Problem: *Draw a circle through two given points.*

B

A

Solution:

1. Draw $\overline{AB}$.

2. Construct the perpendicular bisector of $\overline{AB}$.

3. Choose a point on the perpendicular bisector and label it Q.

4. Draw the circle with center Q and radius $\overline{QA}$.

5. Does the circle pass through B? _ _ _ _ _

6. Draw another circle through points A and B.

7. How many circles can you draw through points A and B? _ _ _ _ _

8. A circle can _ _ _ _ _ _ _ _ _ _ _ _ _ _ _ be drawn through two points.

   (a) always (b) sometimes (c) never

Problem: *Draw a circle through three given points.*

B •

A •

• C

Solution:

1. Draw a circle through points B and C.
2. Does your circle pass through A? _ _ _ _ _
3. Draw $\overline{AB}$ and $\overline{AC}$.
4. Construct the perpendicular bisector of $\overline{AB}$.
5. Construct the perpendicular bisector of $\overline{AC}$.
   Label as O the points where the perpendicular bisectors meet.
6. Draw the circle with center O and radius $\overline{OA}$.
7. Does the circle pass through A, B, and C? _ _ _ _ _

1. Draw a circle through the three points.

2. Draw a circle through the three points.

3. A circle can _________________ be drawn through three points.

(a) always (b) sometimes (c) never

Problem: *Draw a circle through four points.*

1. Try to draw a circle through the four points.

2. Try to draw a circle through the four points below.

3. A circle can _ be drawn through four points.

(a) always (b) sometimes (c) never

Review

1. Circumscribe a circle about the triangle.

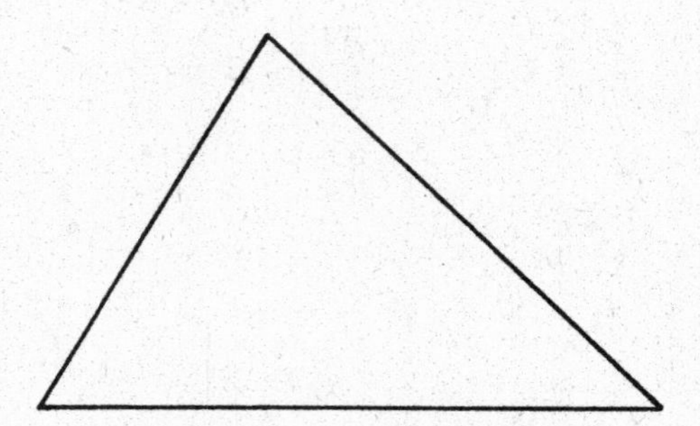

2. Find the center of the circle.

3. Draw a circle through the two given points.

1. Bisect angle ABC.

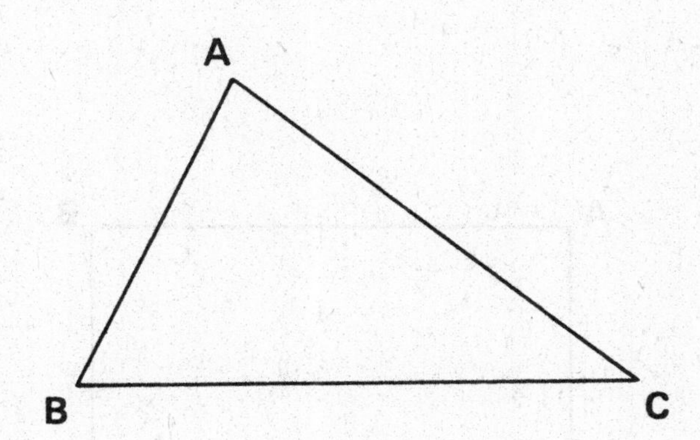

2. Bisect angle ACB.

3. Bisect angle BAC.

4. Label as X the point where the bisectors intersect.

5. Compare $\overline{XA}$, $\overline{XC}$, and $\overline{XB}$.

   Are they congruent? _ _ _ _ _

6. Do you think the circle with center X passing through A will pass through B and C? _ _ _ _ _

7. Draw the circle with center X passing through A.

8. Is this circle the circumscribed circle of triangle ABC? _ _ _ _ _

Circumscribing a Circle about a Polygon

Problem: *Circumscribe a circle about a square.*

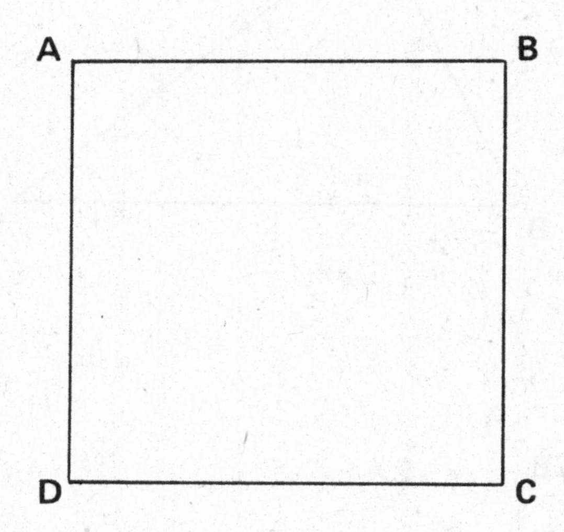

Solution:

1. Construct the perpendicular bisector of $\overline{AB}$.
2. Construct the perpendicular bisector of $\overline{BC}$.
3. Construct the perpendicular bisector of $\overline{CD}$.
4. Construct the perpendicular bisector of $\overline{AD}$.
5. Do the perpendicular bisectors intersect in one point? _ _ _ _ _

   Label the point of intersection P.
6. Draw the circle with center P passing through A.

   Does it circumscribe ABCD? _ _ _ _ _

1. **Circumscribe a circle about the square.**

2. **Circumscribe a circle about the triangle.**

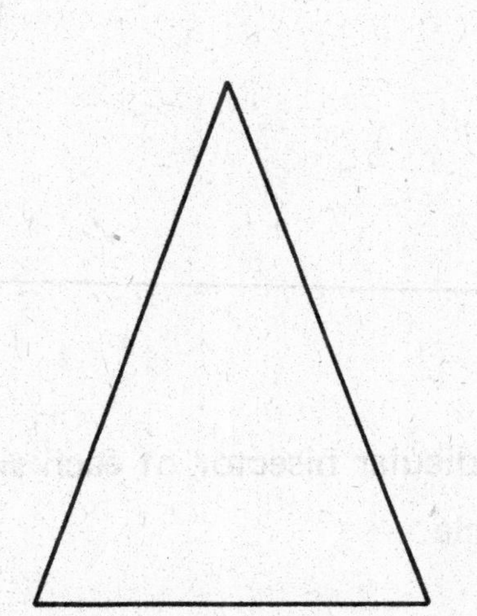

1. Do you think you can circumscribe a circle about a rectangle?

_ _ _ _ _ _

2. Construct the perpendicular bisector of each side, and try to draw the circumscribed circle.

3. Do you think you can circumscribe a circle about the quadrilateral below? _ _ _ _ _

4. Construct the perpendicular bisector of each side, and try to draw the circumscribed circle.

5. A circle can _ _ _ _ _ _ _ _ _ _ _ _ _ _ _ _ _ be circumscribed about a quadrilateral.

   (a) always　　(b) sometimes　　(c) never

1. Do you think you can circumscribe a circle about the pentagon?

_ _ _ _ _

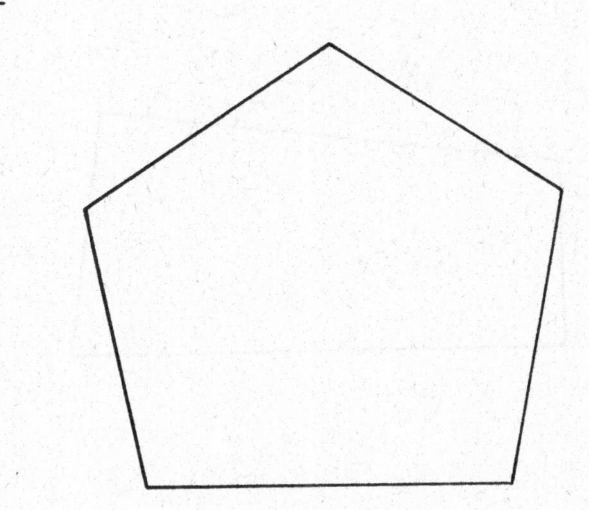

2. Construct the perpendicular bisector of each side and try to draw the circumscribed circle.

3. Do you think you can circumscribe a circle about the parallelogram?

_ _ _ _ _

4. Construct the perpendicular bisector of each side and try to draw the circumscribed circle.

1. Circumscribe a circle about the quadrilateral.

2. Circumscribe a circle about the hexagon.

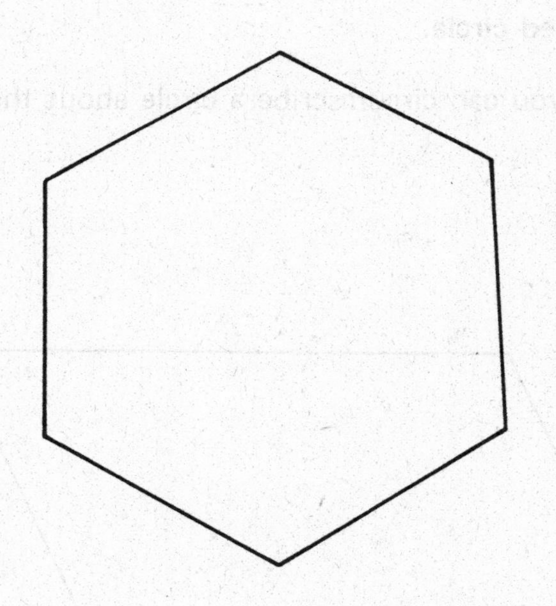

Review

1. Construct a perpendicular to each line through the given point.

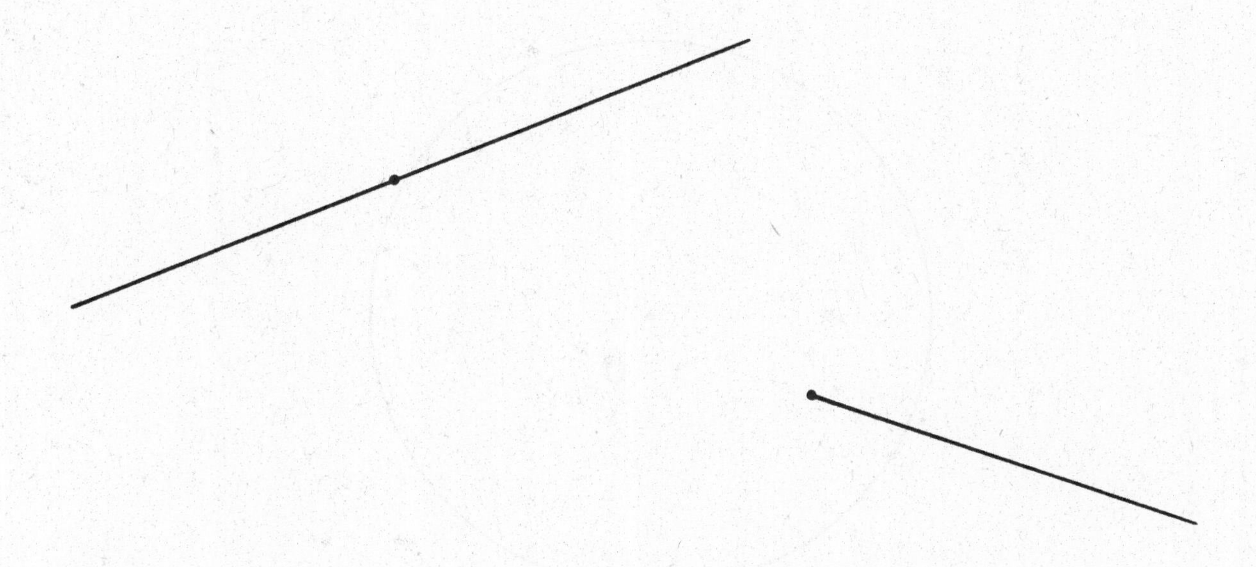

2. Check: Is angle ABC a right angle? _ _ _ _ _

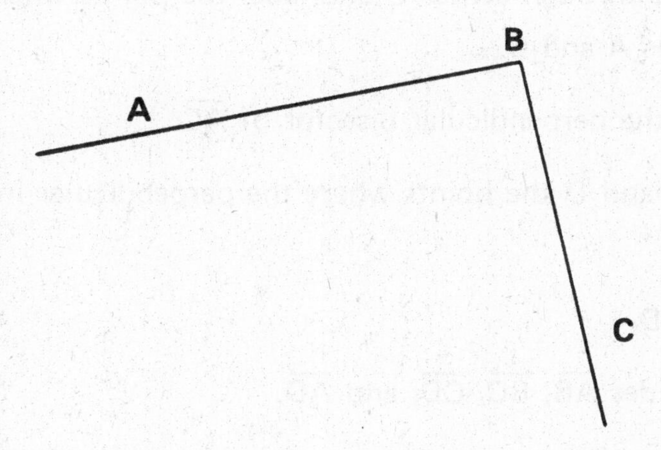

3. Construct a parallel to the given line through the given point.

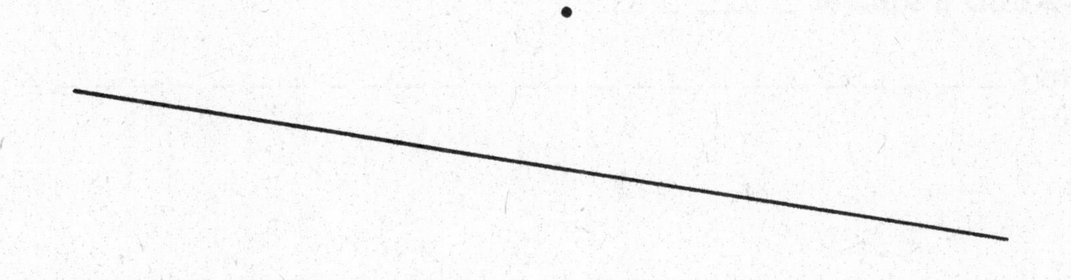

Inscribing a Square in a Circle

Problem: *Inscribe a square in a circle.*

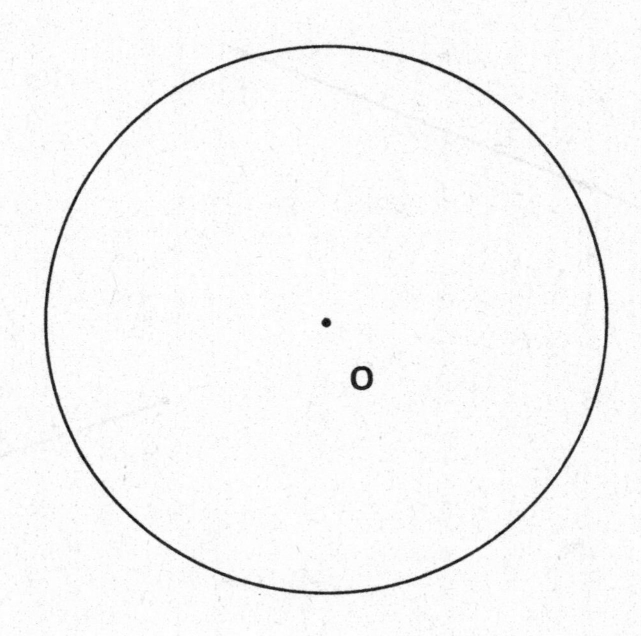

Solution:

1. Draw a line through center O and label the points where it intersects the circle as A and C.
2. Construct the perpendicular bisector of $\overline{AC}$.
3. Label as B and D the points where the perpendicular intersects the circle.
4. Draw ABCD.
5. Compare sides $\overline{AB}$, $\overline{BC}$, $\overline{CD}$, and $\overline{AD}$.

   Are they congruent? _ _ _ _ _
6. Check: Are the angles ABC, BCD, CDA, and BAD right angles? _ _ _ _
7. Is ABCD a square? _ _ _ _ _

   Why? _

1. Inscribe a square in the circle.

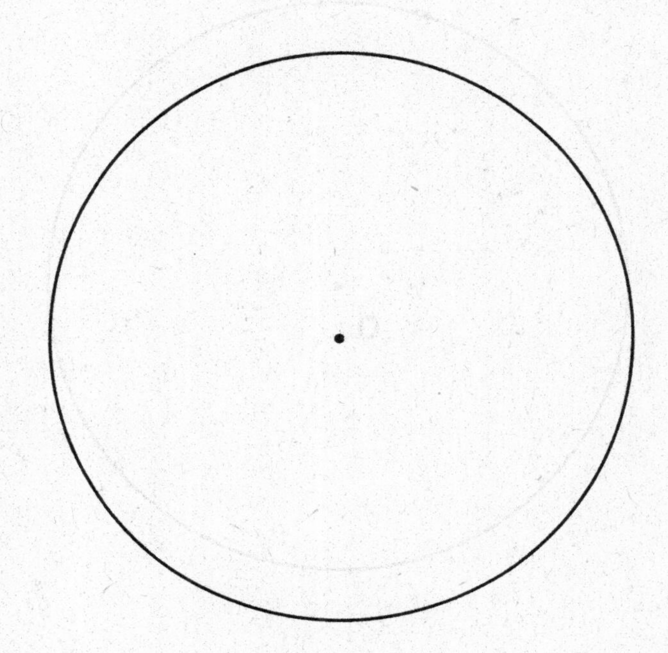

2. Bisect the angles.

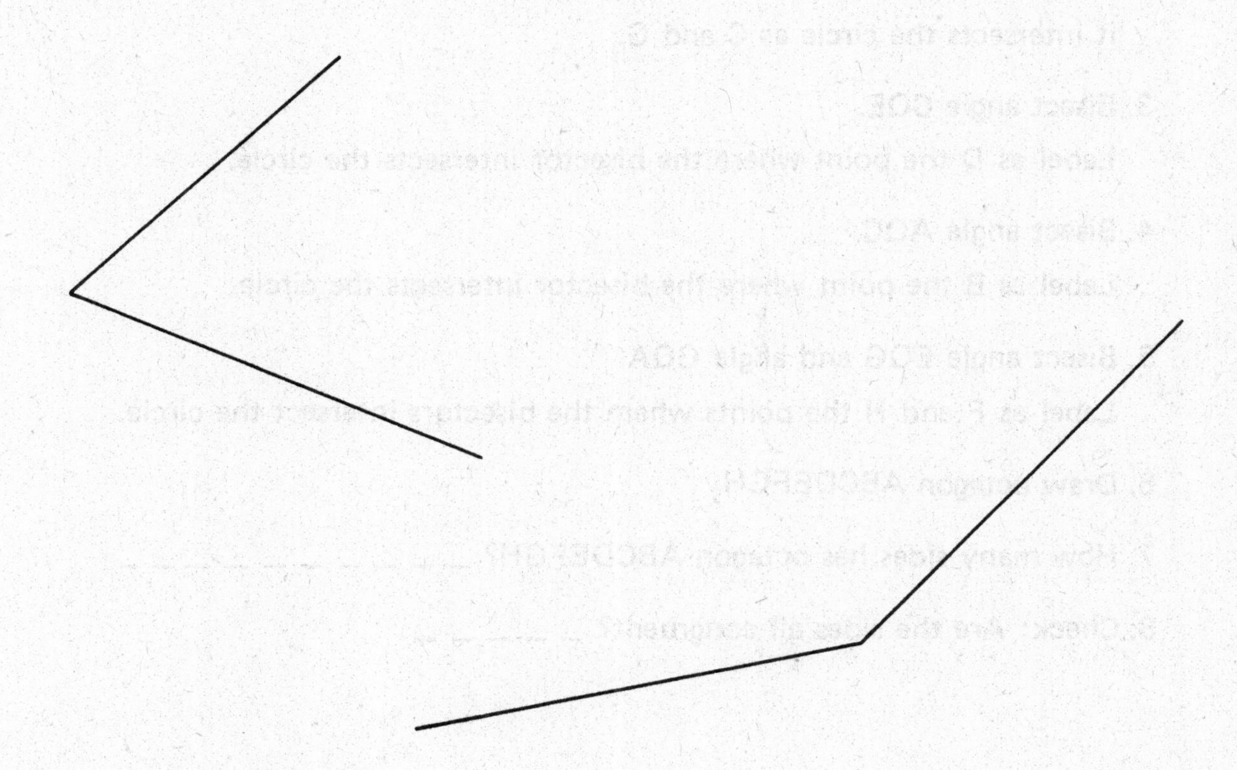

Inscribing a Regular Octagon in a Circle

Problem: *Inscribe a regular octagon in a circle.*

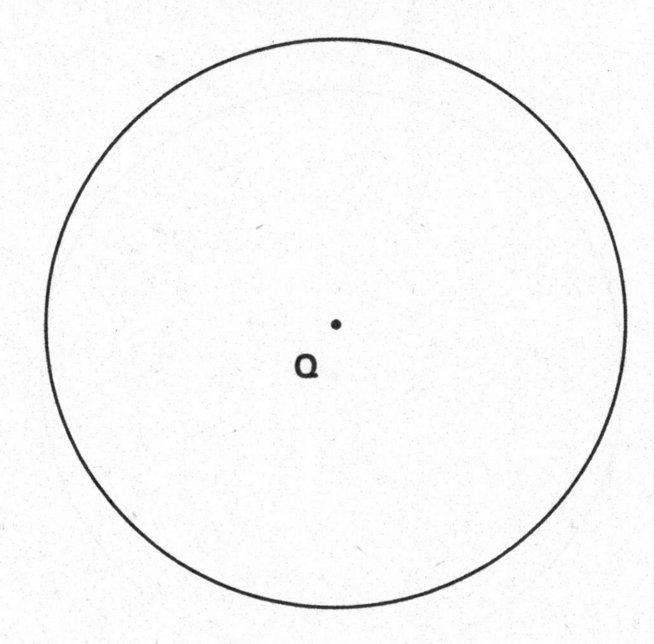

Solution:

1. Draw a line through center Q and label the points where it intersects the circle as A and E.

2. Construct the perpendicular bisector of $\overline{AE}$ and label the points where it intersects the circle as C and G.

3. Bisect angle CQE.
   Label as D the point where the bisector intersects the circle.

4. Bisect angle AQC.
   Label as B the point where the bisector intersects the circle.

5. Bisect angle EQG and angle GQA.
   Label as F and H the points where the bisectors intersect the circle.

6. Draw octagon ABCDEFGH.

7. How many sides has octagon ABCDEFGH? _ _ _ _ _ _ _ _ _ _

8. Check: Are the sides all congruent? _ _ _ _ _

1. Inscribe a regular octagon in the circle.

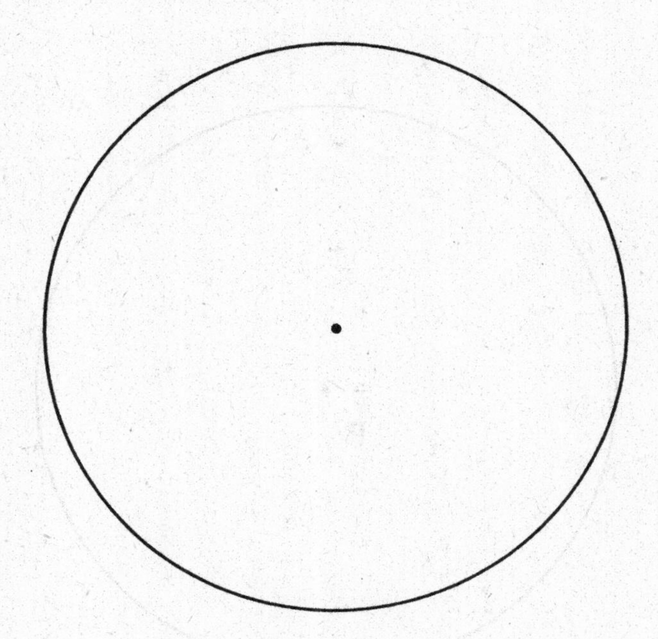

2. Construct an equilateral triangle with $\overline{AB}$ as one side.

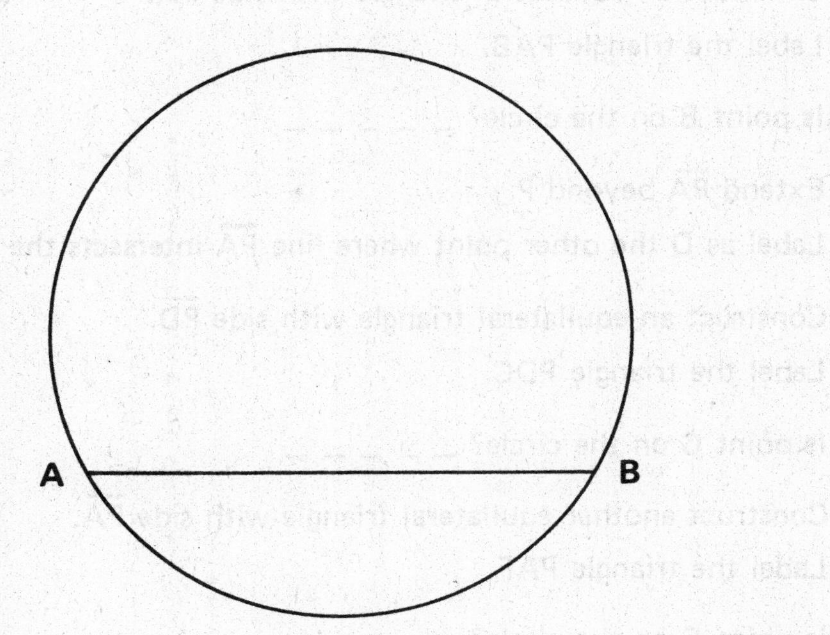

Inscribing a Regular Hexagon in a Circle

Problem: *Inscribe a regular hexagon in the circle.*

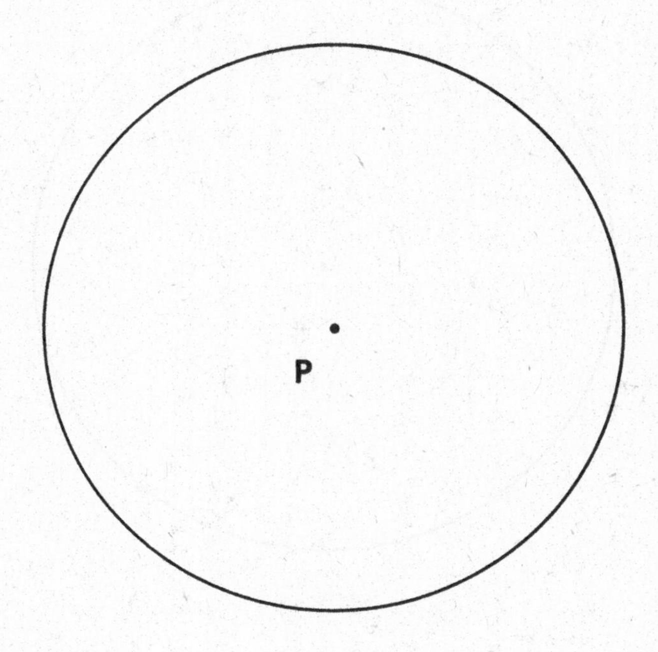

Solution:

1. Draw a radius of the circle and label it $\overline{PA}$.
2. Construct an equilateral triangle with side $\overline{PA}$.
   Label the triangle PAB.
3. Is point B on the circle? _ _ _ _ _
4. Extend $\overline{PA}$ beyond P.
   Label as D the other point where line $\overleftrightarrow{PA}$ intersects the circle.
5. Construct an equilateral triangle with side $\overline{PD}$.
   Label the triangle PDC.
6. Is point C on the circle? _ _ _ _ _
7. Construct another equilateral triangle with side $\overline{PA}$.
   Label the triangle PAF.
8. Is point F on the circle? _ _ _ _ _

Go on to the next page.

9. Construct another equilateral triangle with side $\overline{PD}$.
   Label the triangle PDE.

10. Is point E on the circle? _ _ _ _ _

11. Draw hexagon ABCDEF.

12. How many sides has hexagon ABCDEF? _ _ _ _ _

13. Check: Are all the sides congruent? _ _ _ _ _

1. Inscribe a regular hexagon in circle Q.

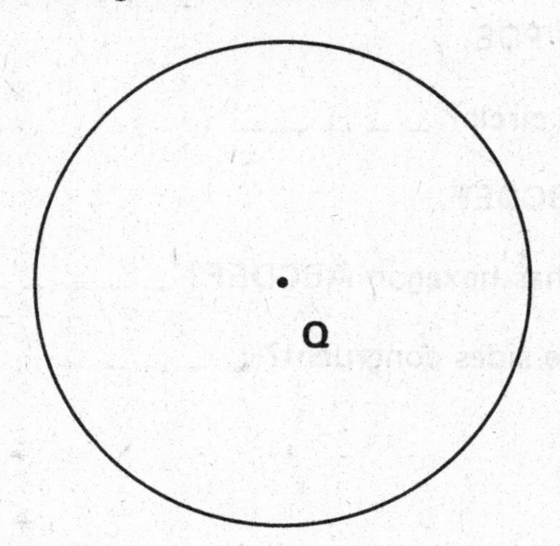

2. Label the hexagon ABCDEF.

3. Compare $\overline{QA}$ and $\overline{CD}$.

   Are they congruent? _ _ _ _ _

4. The sides of a regular hexagon inscribed in a circle are

   _ _ _ _ _ _ _ _ _ _ _ _ _ _ _ the radius of the circle.

   (a) shorter than
   (b) congruent to
   (c) longer than

5. Inscribe a square in the circle.

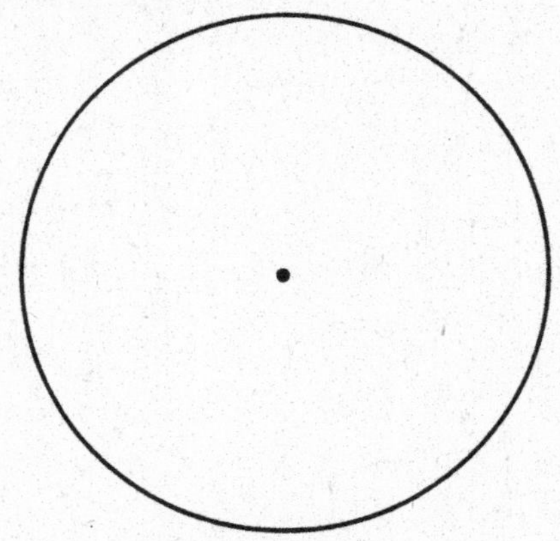

6. Compare a radius and a side of the square.

7. The sides of a square inscribed in a circle are _ _ _ _ _ _ _ _ _ _ _ the radius of the circle.

   (a) shorter than
   (b) congruent to
   (c) longer than

Inscribe in the circle a regular polygon with 12 sides.

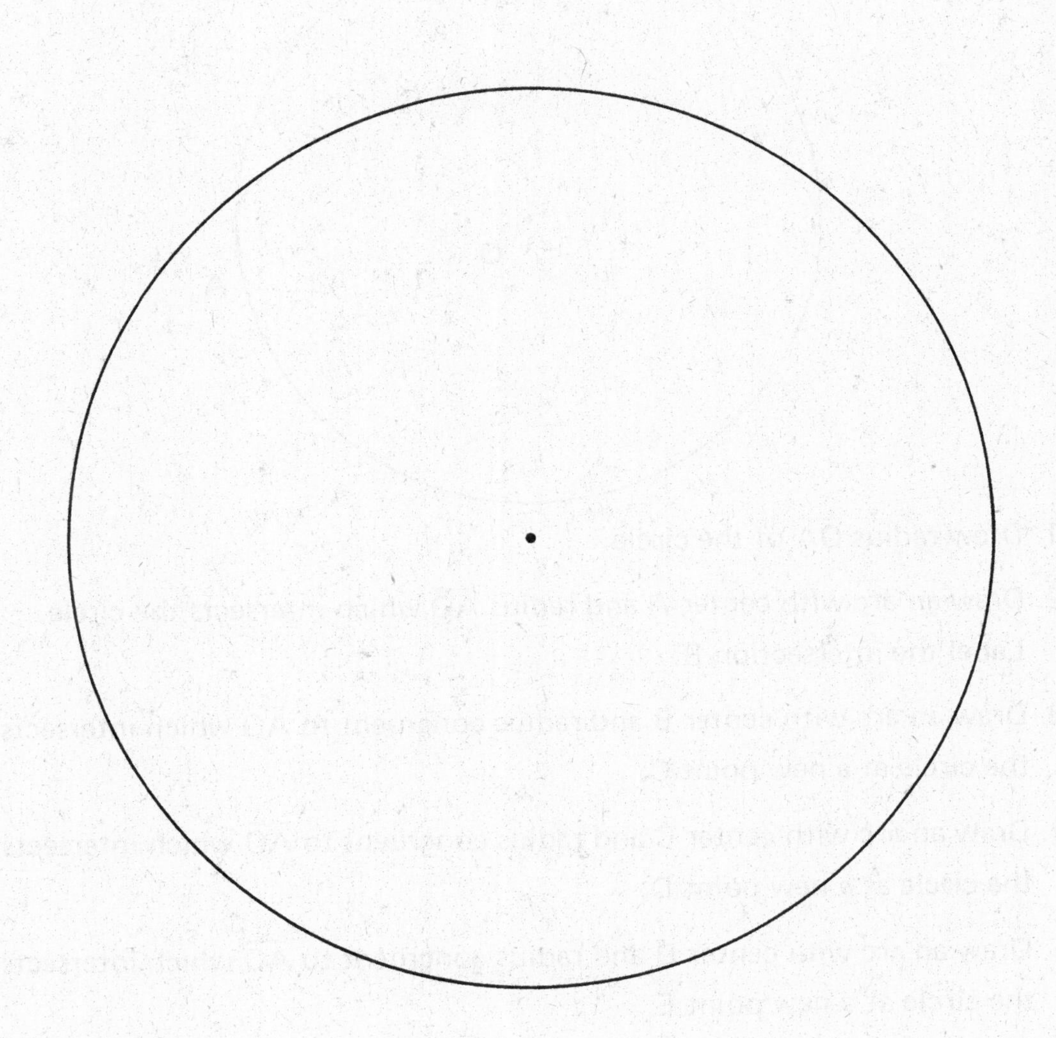

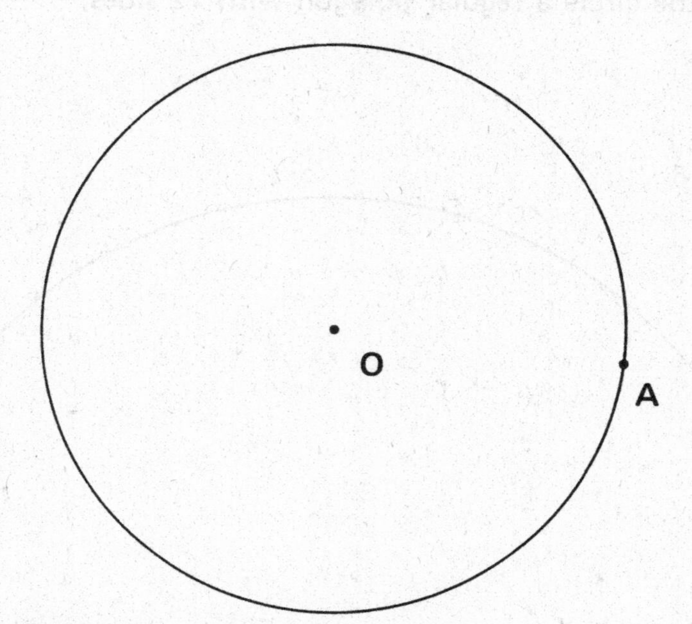

1. Draw radius $\overline{OA}$ of the circle.

2. Draw an arc with center A and radius $\overline{AO}$ which intersects the circle. Label the intersection B.

3. Draw an arc with center B and radius congruent to $\overline{AO}$ which intersects the circle at a new point C.

4. Draw an arc with center C and radius congruent to $\overline{AO}$ which intersects the circle at a new point D.

5. Draw an arc with center D and radius congruent to $\overline{AO}$ which intersects the circle at a new point E.

6. Draw an arc with center E and radius congruent to $\overline{AO}$ which intersects the circle at a new point F.

7. Draw an arc with center E and radius congruent to $\overline{AO}$ which intersects the circle at a new point.

   Does this arc intersect the circle at A? _ _ _ _ _

8. Draw polygon ABCDEF.

9. What kind of polygon is ABCDEF? _ _ _ _ _

10. Compare the sides of the polygon.

    Are they all congruent? _ _ _ _ _

Problem: *Inscribe an equilateral triangle in a circle.*

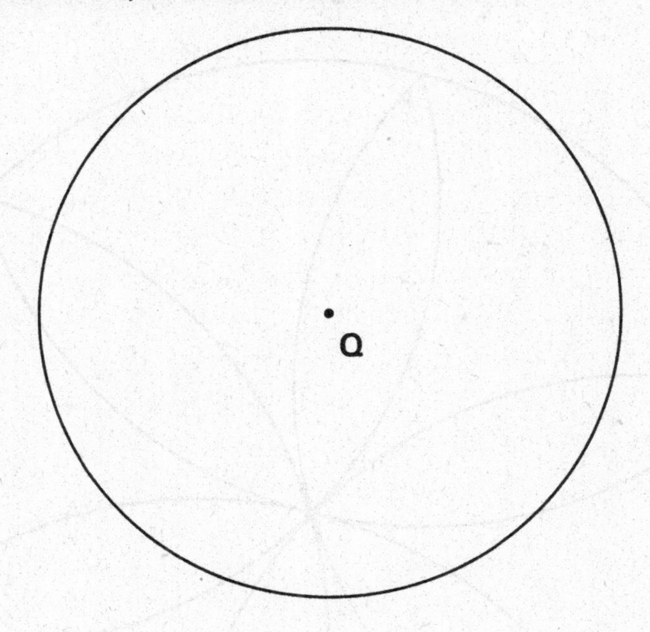

Solution:

1. Draw a line through center Q.
   Label as A and B the points where the line intersects the circle.
2. Draw two arcs with center A and radius $\overline{AQ}$ which intersect the circle.
   Label as C and D the points of intersection.
3. Draw triangle BCD.
4. Check: Are sides $\overline{BC}$, $\overline{CD}$, and $\overline{BD}$ all congruent? _ _ _ _ _
5. Check: Are angles BCD, CED, and BDC all congruent? _ _ _ _ _
6. Inscribe an equilateral triangle in circle O.

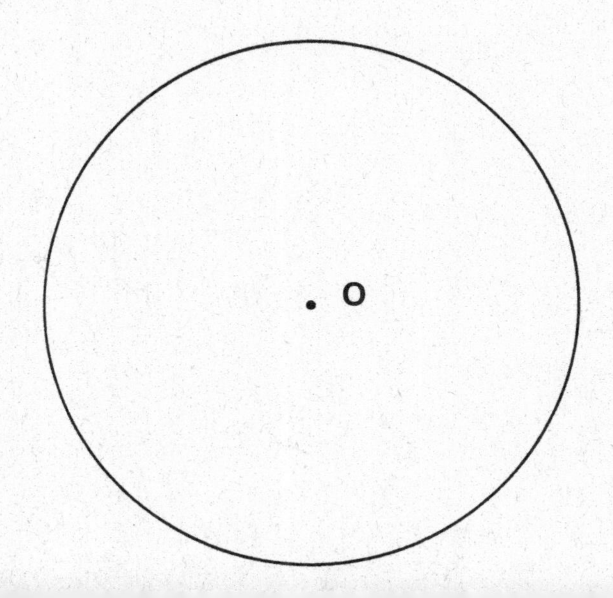

Use your compass to copy this figure.

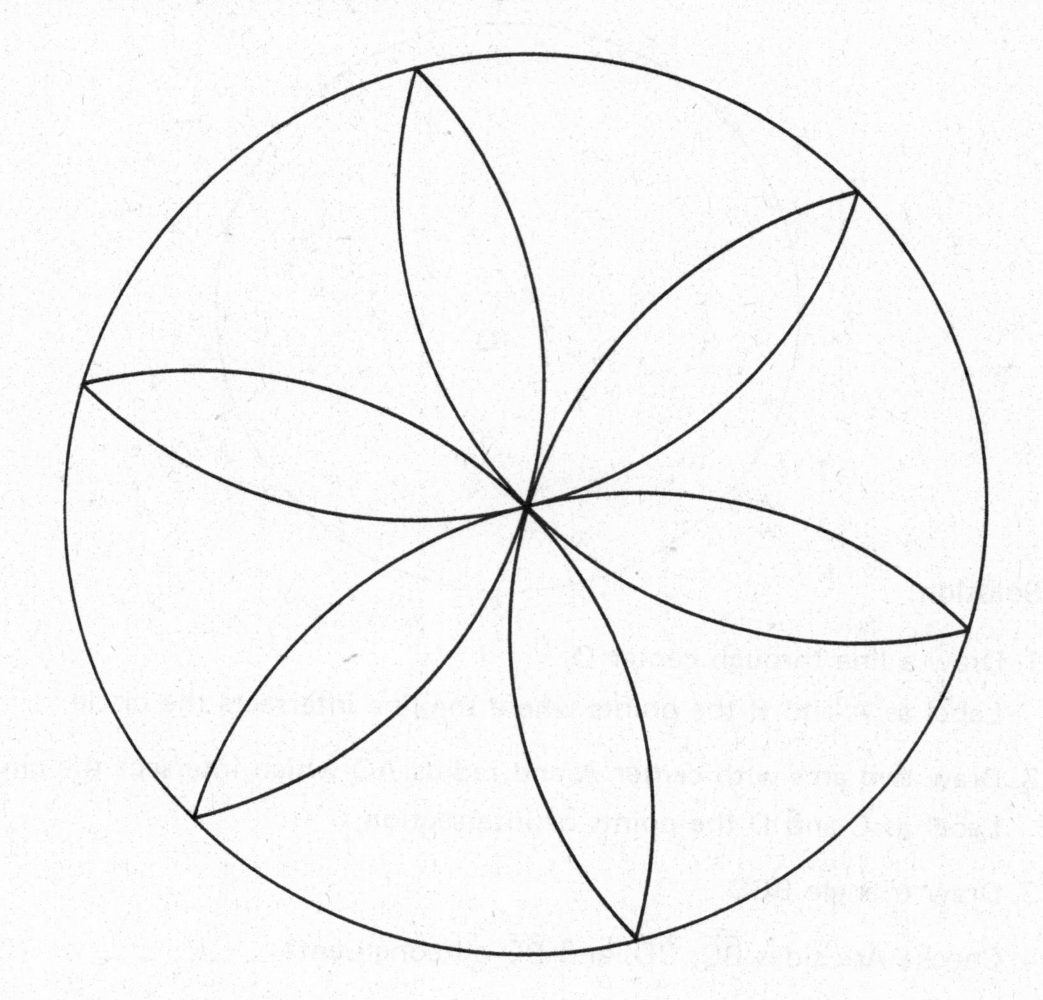

Inscribing a Regular Pentagon in a Circle

Problem: *Inscribe a regular pentagon in a circle.*

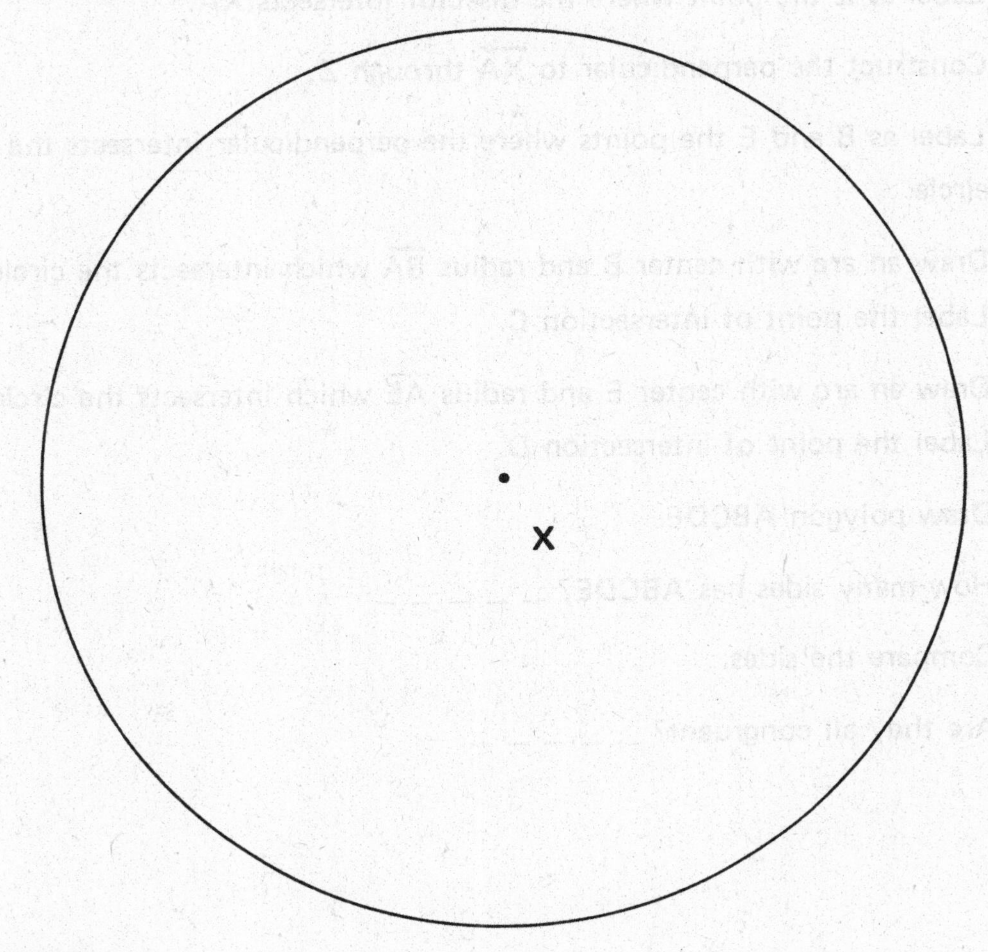

Solution:

1. Draw a line through center X of the circle.
   Label as A one of the points where it intersects the circle.
2. Construct the perpendicular to $\overleftrightarrow{AX}$ through point X.
   Label as Y one of the points where it intersects the circle.
3. Bisect $\overline{YX}$.
   Label the midpoint M.
4. Draw $\overline{MA}$.

Go on to the next page.

5. Bisect angle AMX.

   Label as Z the point where the bisector intersects $\overleftrightarrow{XA}$.

6. Construct the perpendicular to $\overleftrightarrow{XA}$ through Z.

7. Label as B and E the points where the perpendicular intersects the circle.

8. Draw an arc with center B and radius $\overline{BA}$ which intersects the circle.

   Label the point of intersection C.

9. Draw an arc with center E and radius $\overline{AE}$ which intersects the circle.

   Label the point of intersection D.

10. Draw polygon ABCDE.

11. How many sides has ABCDE? _____

12. Compare the sides.

    Are they all congruent? _____

Inscribe a regular <u>decagon</u> in the circle. (A decagon has 10 sides.)

Inscribing a Circle in a Triangle

Problem: *Inscribe a circle in a triangle.*

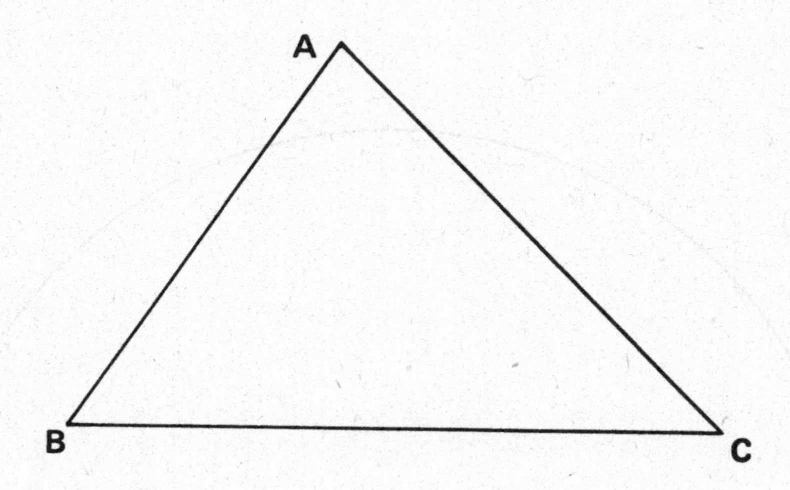

Solution:

1. Bisect angle ABC, angle ACB, and angle BAC of the triangle.

2. Do the bisectors meet in one point? _ _ _ _ _

   Label the point of intersection Q.

3. Construct the perpendicular from Q to line $\overleftrightarrow{BC}$.

4. Label as R the point where the perpendicular intersects $\overleftrightarrow{BC}$.

5. Draw the circle with center Q passing through R.

6. Is the circle tangent to side $\overline{BC}$? _ _ _ _ _

   Is the circle tangent to side $\overline{AC}$? _ _ _ _ _

   Is the circle tangent to side $\overline{AB}$? _ _ _ _ _

   The circle is inscribed in triangle ABC.

1. Inscribe a circle in the triangle.

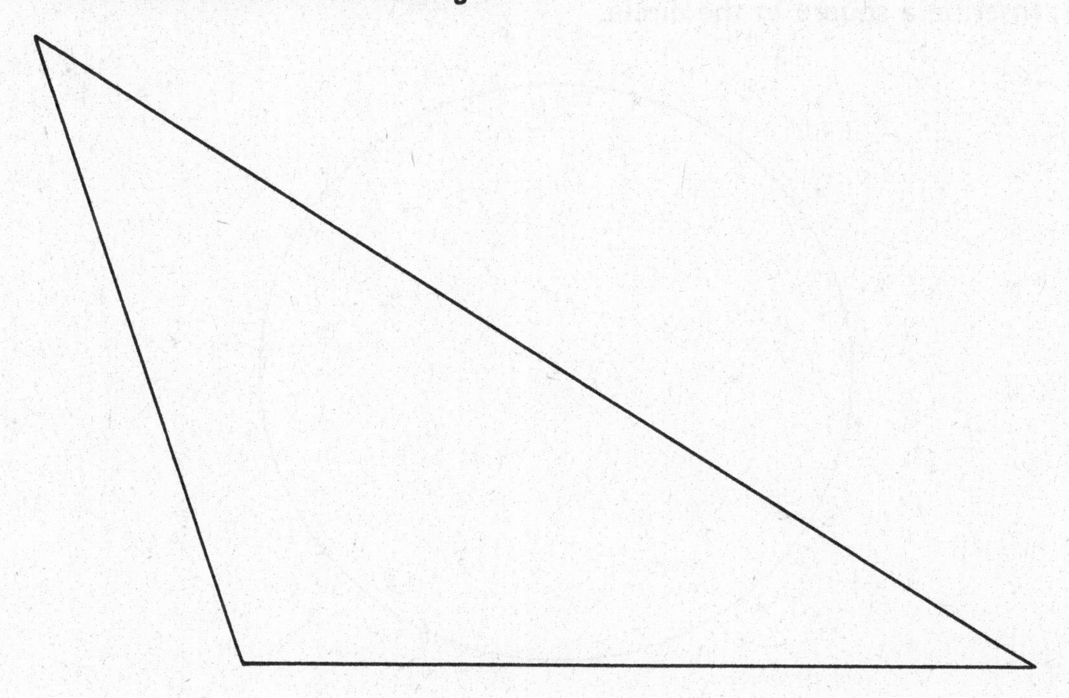

2. Construct an equilateral triangle with side $\overline{AB}$.

A ———————————————————— B

3. Inscribe a circle in the equilateral triangle.

4. A circle can _ _ _ _ _ _ _ _ _ _ _ _ _ _ _ be inscribed in a triangle.

(a) always (b) sometimes (c) never

Review

1. Inscribe a square in the circle.

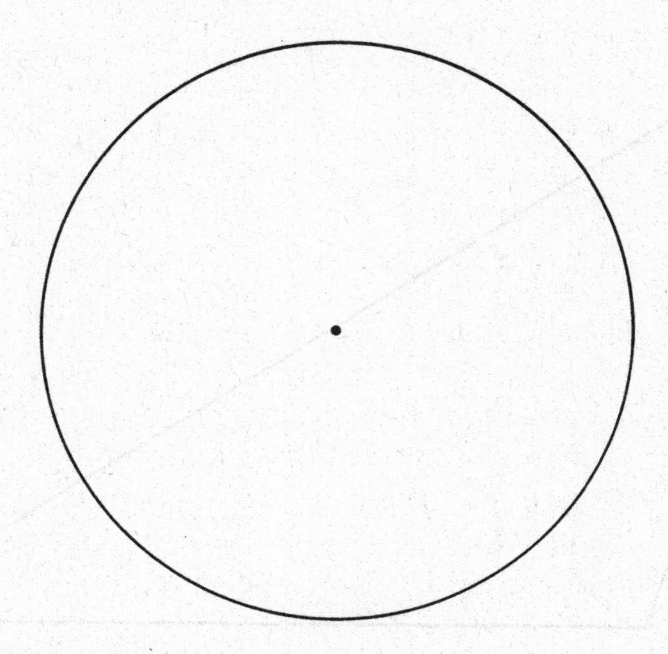

2. Inscribe a circle in the square above.

3. Circumscribe a circle about square ABCD.

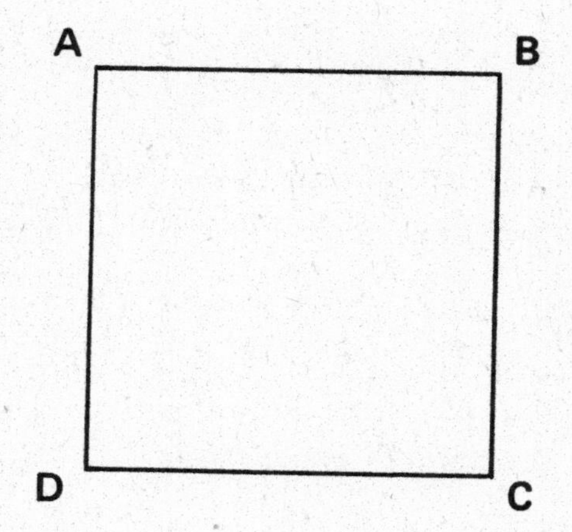

Inscribing a Circle in a Regular Polygon

Problem: *Inscribe a circle in a rhombus.*

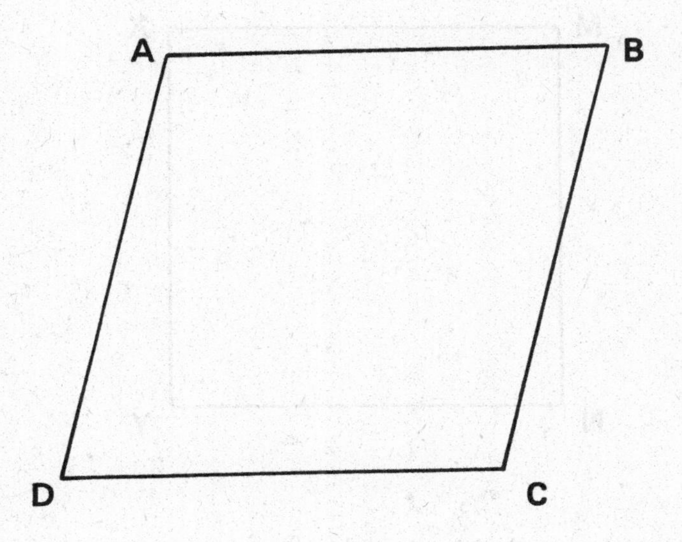

Solution:

1. Bisect each angle ADC, DCB, CBA, and BAD of the rhombus.
2. Do the bisectors meet in a point? _ _ _ _ _

   Label the point of intersection O.
3. Construct the perpendicular from O to $\overleftrightarrow{DC}$.
4. Label as P the point where the perpendicular intersects $\overleftrightarrow{DC}$.
5. Draw the circle with center O passing through P.
6. Is this circle tangent to each side? _ _ _ _ _
7. Is the circle inscribed in rhombus ABCD? _ _ _ _ _

1. Do you think you can inscribe a circle in square MXYN? _ _ _ _ _

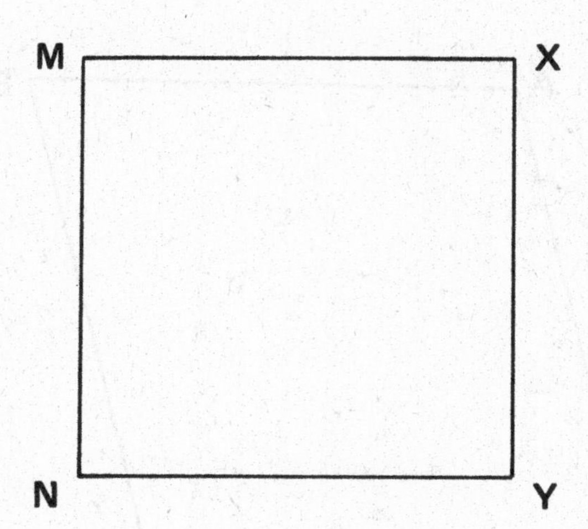

2. Construct the bisector of each angle and try to inscribe a circle in the square.

3. Do you think you can inscribe a circle in rectangle ABCD? _ _ _ _

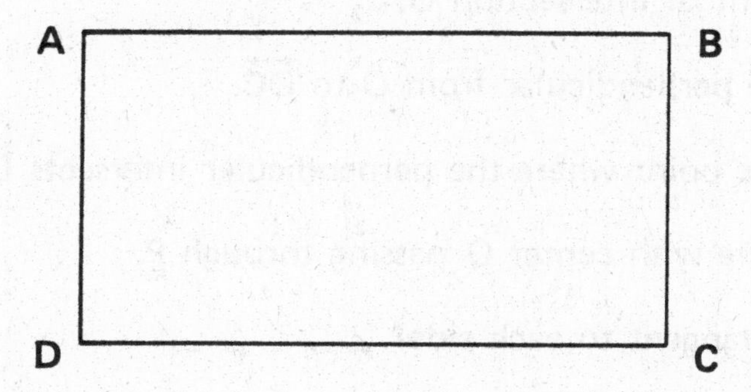

4. Construct the bisector of each angle and try to inscribe a circle in the rectangle.

5. A circle can _ _ _ _ _ _ _ _ _ _ _ _ _ _ _ _ _ _ be inscribed in a quadrilateral.

   (a) always (b) sometimes (c) never

1. Inscribe a circle in the pentagon.

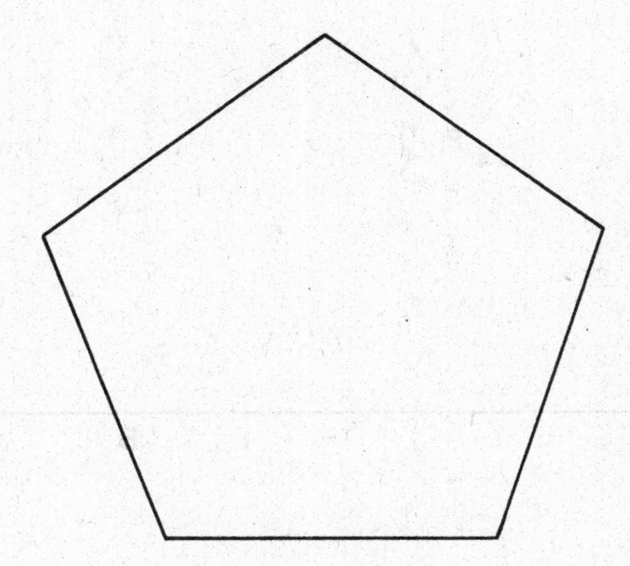

2. Circumscribe a circle about the pentagon.

3. Inscribe a circle in the hexagon.

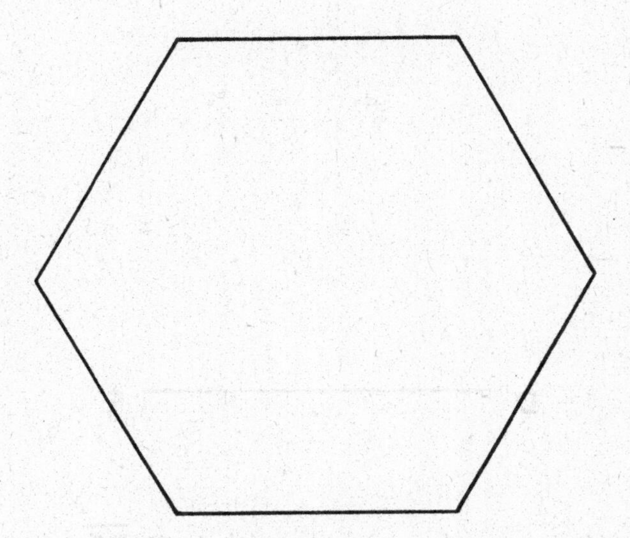

4. Circumscribe a circle about the hexagon.

Practice Test

1. Construct a perpendicular to $\overleftrightarrow{AB}$ through P.

P

A B

2. Construct a parallel to $\overleftrightarrow{AB}$ through P.

3. Construct an isosceles triangle with base $\overline{DE}$.

D E

4. Construct another isosceles triangle with base $\overline{DE}$.

5. Construct a perpendicular to the line through the given point.

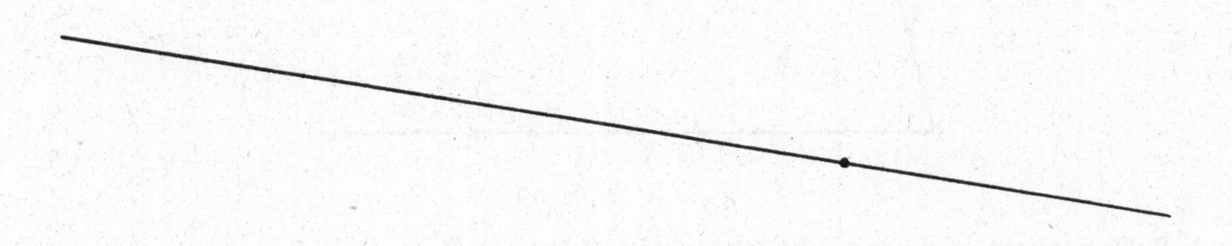

6. Construct a rectangle with sides congruent to $\overline{AB}$ and $\overline{CD}$.

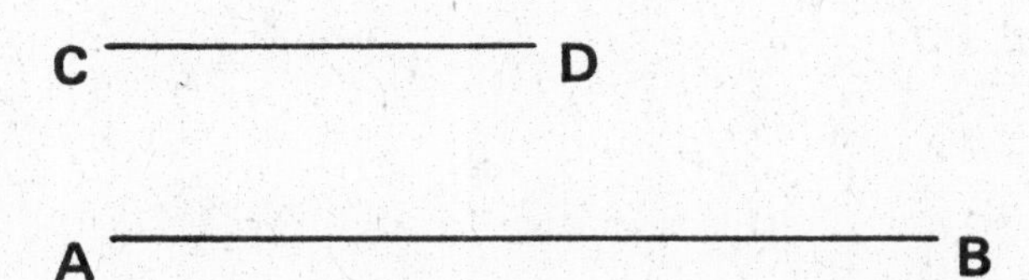

7. Construct the rest of the parallelogram with the two given sides.

8. Construct a square with the given side.

9. Bisect the angle.

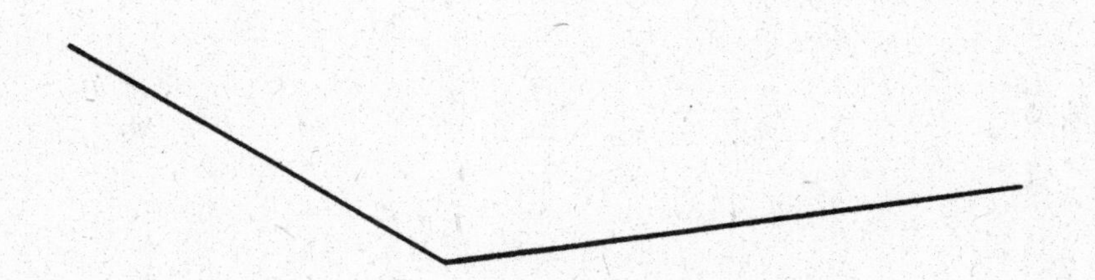

10. Construct a circle tangent to line $\overleftrightarrow{AB}$ at point A.

11. Construct the perpendicular bisector of each segment.

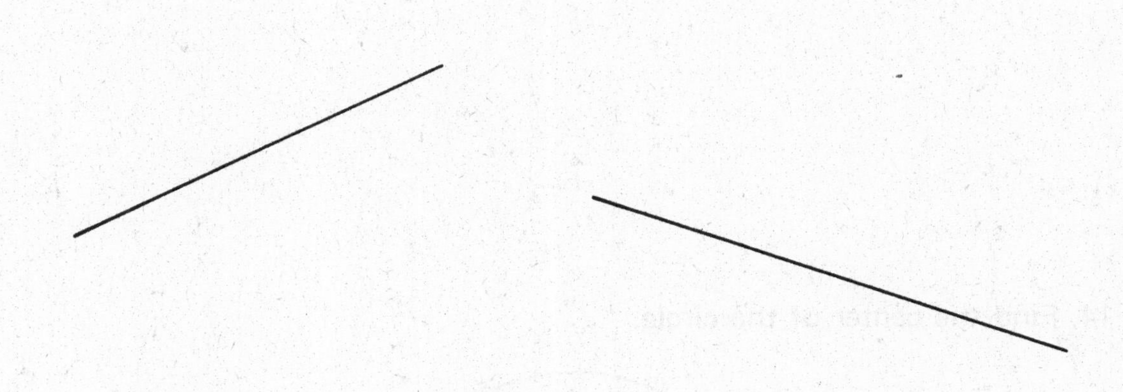

12. Construct a line tangent to the circle through the given point.

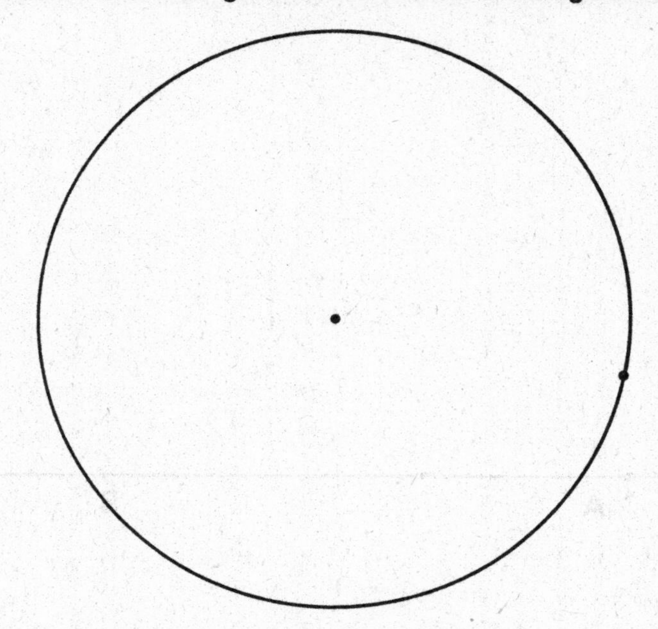

13. Circumscribe a circle about triangle ABC.

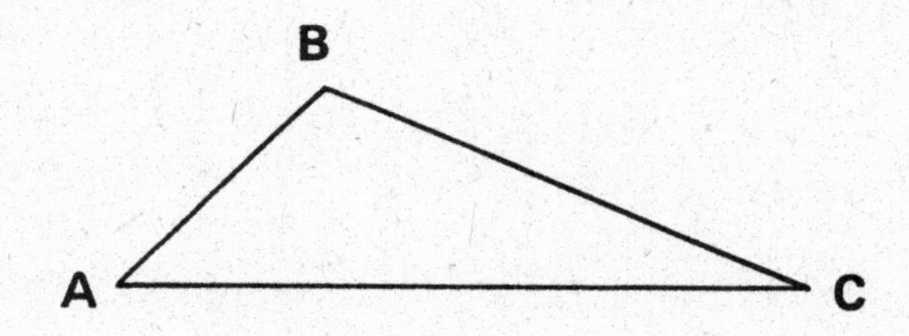

14. Find the center of the circle.

15. Construct two circles tangent to the given circle with center C at point P.

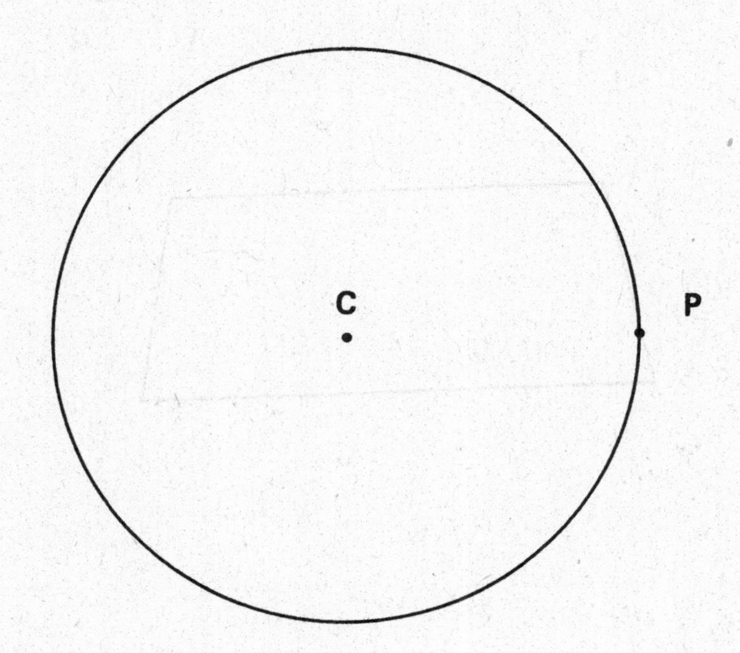

16. Bisect the angle.

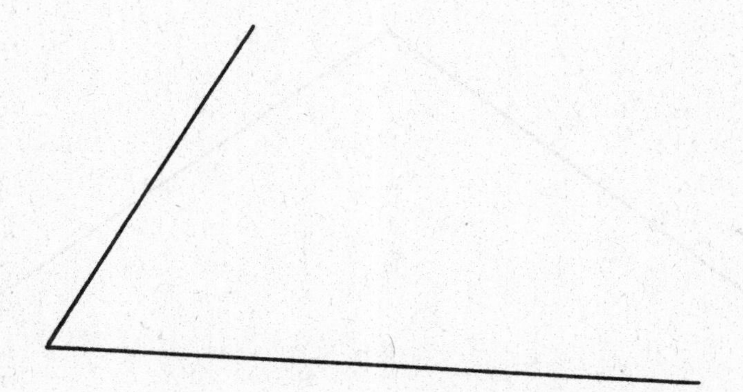

17. Inscribe a circle in the rhombus.

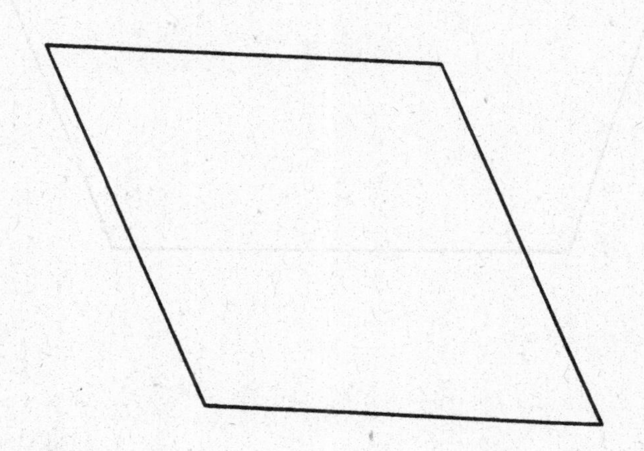

18. Circumscribe a circle about the quadrilateral.

19. Inscribe a circle in the pentagon.

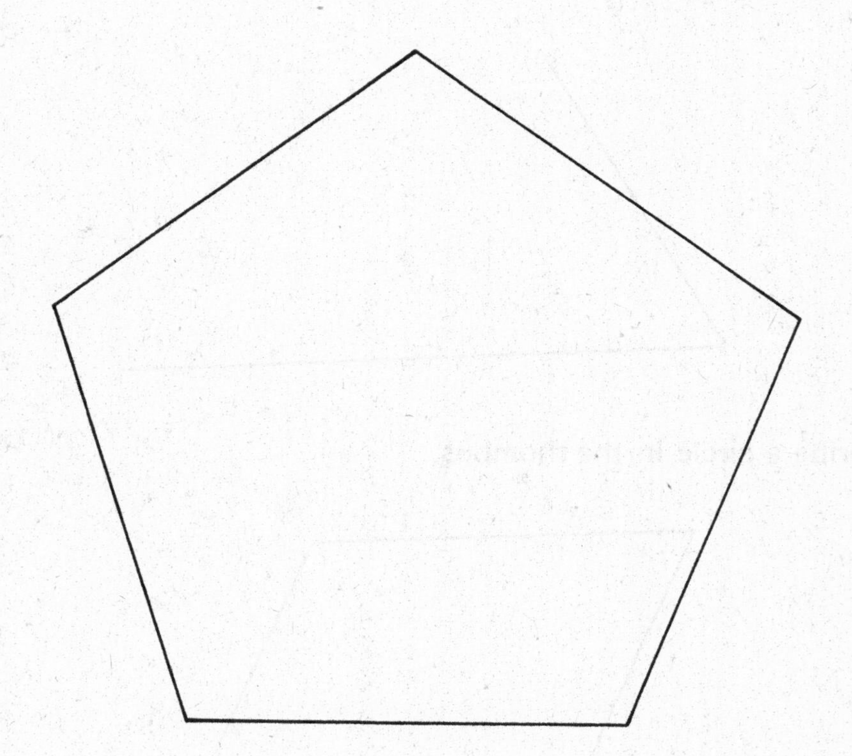

20. Draw a circle through points A, B, and C.

B

A C

21. Inscribe a regular hexagon in the circle.

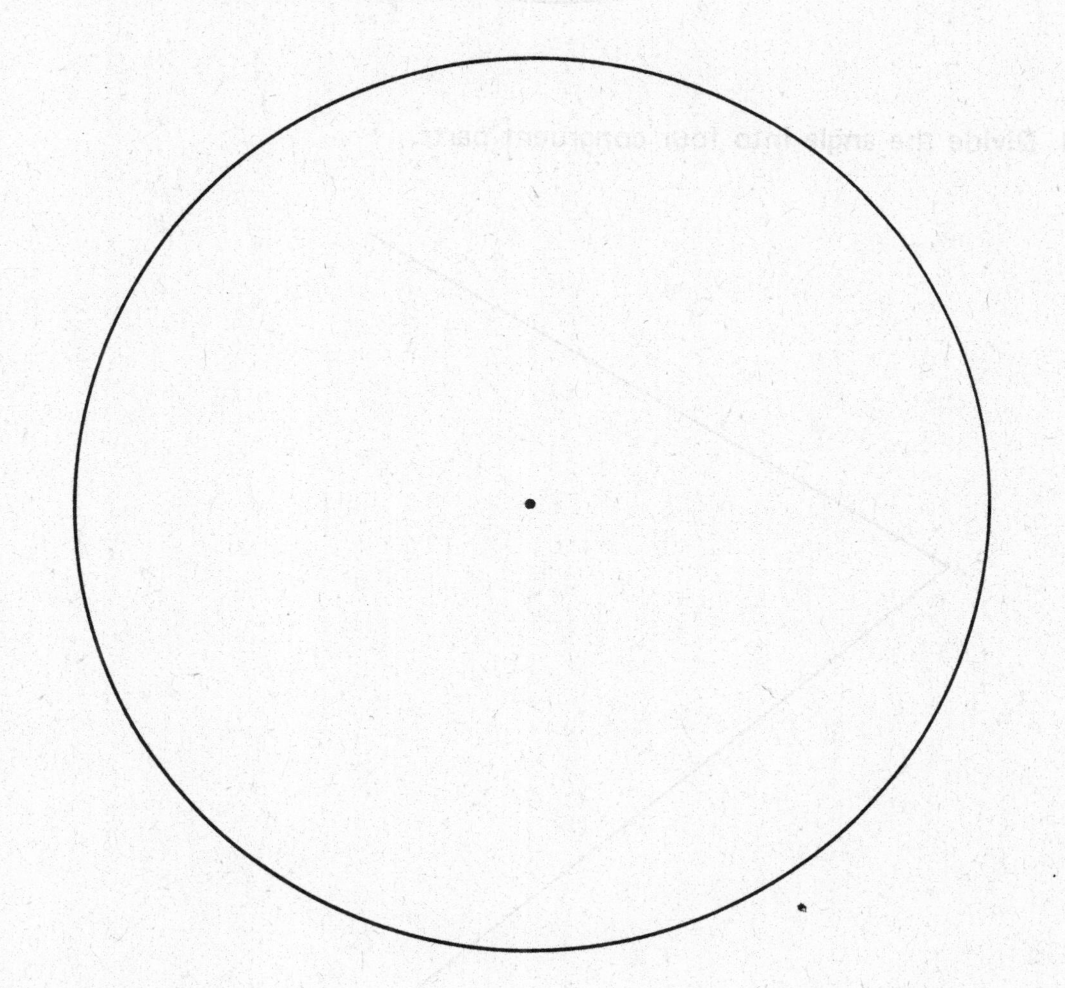

22. Inscribe a regular octagon in the circle.

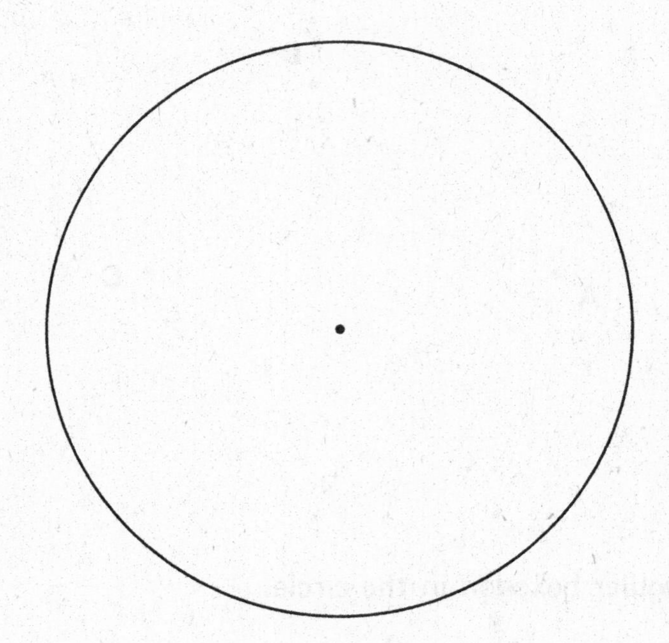

23. Divide the angle into four congruent parts.

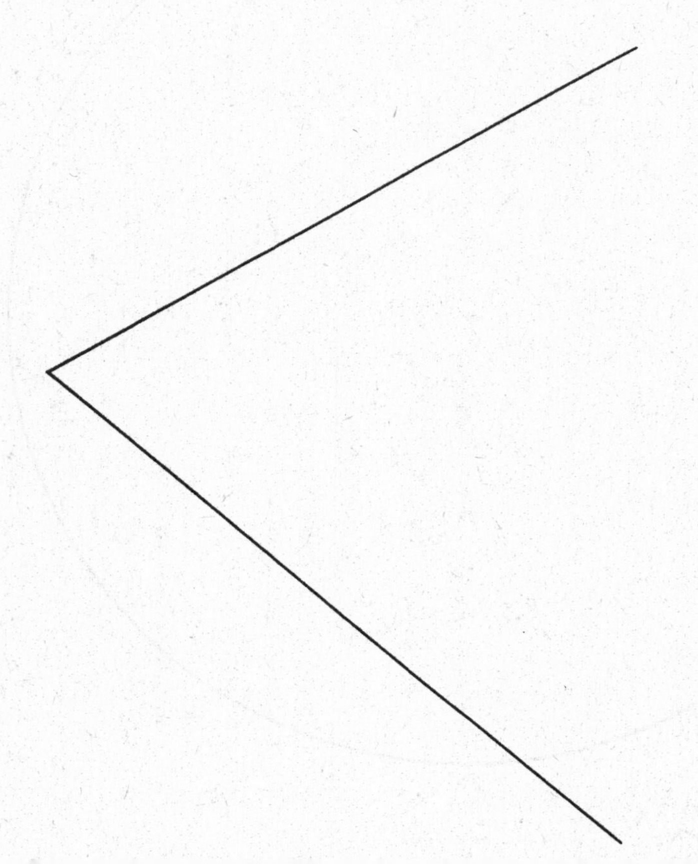

KEY TO GEOMETRY, Books 1-8

INDEX OF UNDERLINED TERMS

(Numbers indicate book and page.)